SpringerBriefs in Molecular Science

Chemistry of Foods

Series Editor

Salvatore Parisi, Lourdes Matha Institute of Hotel Management and Catering Technology, Thiruvananthapuram, Kerala, India

The series Springer Briefs in Molecular Science: Chemistry of Foods presents compact topical volumes in the area of food chemistry. The series has a clear focus on the chemistry and chemical aspects of foods, topics such as the physics or biology of foods are not part of its scope. The Briefs volumes in the series aim at presenting chemical background information or an introduction and clear-cut overview on the chemistry related to specific topics in this area. Typical topics thus include:

- Compound classes in foods—their chemistry and properties with respect to the foods (e.g. sugars, proteins, fats, minerals, …)
- Contaminants and additives in foods—their chemistry and chemical transformations
- Chemical analysis and monitoring of foods
- Chemical transformations in foods, evolution and alterations of chemicals in foods, interactions between food and its packaging materials, chemical aspects of the food production processes
- Chemistry and the food industry—from safety protocols to modern food production

The treated subjects will particularly appeal to professionals and researchers concerned with food chemistry. Many volume topics address professionals and current problems in the food industry, but will also be interesting for readers generally concerned with the chemistry of foods. With the unique format and character of SpringerBriefs (50 to 125 pages), the volumes are compact and easily digestible. Briefs allow authors to present their ideas and readers to absorb them with minimal time investment. Briefs will be published as part of Springer's eBook collection, with millions of users worldwide. In addition, Briefs will be available for individual print and electronic purchase. Briefs are characterized by fast, global electronic dissemination, standard publishing contracts, easy-to-use manuscript preparation and formatting guidelines, and expedited production schedules.

Both solicited and unsolicited manuscripts focusing on food chemistry are considered for publication in this series. Submitted manuscripts will be reviewed and decided by the series editor, Prof. Dr. Salvatore Parisi.

To submit a proposal or request further information, please contact Dr. Sofia Costa, Publishing Editor, via sofia.costa@springer.com or Prof. Dr. Salvatore Parisi, Book Series Editor, via drparisi@inwind.it or drsalparisi5@gmail.com

More information about this subseries at http://www.springer.com/series/11853

Ramesh Kumar Sharma · Maria Micali ·
Bhupendra Kumar Rana · Alessandra Pellerito ·
Rajeev K. Singla

Indian Herbal Medicines

Antioxidant and Antimicrobial Properties

Ramesh Kumar Sharma
Food safety consultant
Tilam Sangh Rajasthan
Bikaner, Rajasthan, India

Bhupendra Kumar Rana
Quality and Accreditation Institute
Noida, Uttar Pradesh, India

Rajeev K. Singla
Institutes for Systems Genetics
West China Hospital, Sichuan University
Chengdu, Sichuan, China

Maria Micali
Food safety consultant, Messina, Italy

Alessandra Pellerito
Food safety consultant, Palermo, Italy

ISSN 2191-5407 ISSN 2191-5415 (electronic)
SpringerBriefs in Molecular Science
ISSN 2199-689X ISSN 2199-7209 (electronic)
Chemistry of Foods
ISBN 978-3-030-80917-1 ISBN 978-3-030-80918-8 (eBook)
https://doi.org/10.1007/978-3-030-80918-8

This Springer imprint is published by the registered company Springer Nature Switzerland AG
The registered company address is: Gewerbestrasse 11, 6330 Cham, Switzerland

Contents

About the Authors

Ramesh Kumar Sharma is currently Food Safety Consultant and Scientific Writer 2000 onwards, having worked in past as Chemist in different roles including: Fellow, Science Education Centre, University of Rajasthan, Jaipur 1978–1979; Chief Investigator, Play Material Project—UNICEF project. *Toxins and Contaminants in Indian Food Products* is the title of a book, written by him in co-authorship with Salvatore Parisi, published by Springer.

Maria Micali is an experienced Author in the field of food science and technology, with particular focus in chemistry, microbiology and hygiene. She obtained a Ph.D. in food hygiene from the University of Messina, Italy. She is also Lecturer in different sectors, including professional training. Her published works include *The Chemistry of Thermal Food Processing Procedures* (2016) and *Traceability in the Cheesemaking Field. The Regulatory Ambit and Practical Solutions* (2016).

Bhupendra Kumar Rana is a seasoned quality professional, extensively worked in healthcare quality and improvement. He obtained a Ph.D. in Biochemistry from Banaras Hindu University, Varanasi, India. From the past 20 years, he has been pioneer in accreditation standards, currently being CEO of Quality & Accreditation Institute (QAI) and International Expert/Consultant for the World Health Organization, the World Bank, Asian Development Bank, USAID and several Ministries of Health. He has published 12 research papers and several chapters in books on antimicrobial activity of plant products and healthcare quality.

Alessandra Pellerito is a Biologist graduated at the University of Bologna, Italy (2013), with full marks (110/100 cum laude) after the initial B.Sc. in Biology (Palermo, Italy). After a short period spent in the UK, she moved to Germany (Magdeburg). At present, she works as Food Consultant in the private sector (Italy). Her first articles on food chemistry have been published by the *Journal of AOAC International*. Her first book is *Food Sharing—Chemical Evaluation of Durable Foods* (Springer).

Rajeev K. Singla works as Assistant Researcher in Institutes for Systems Genetics, West China Hospital of Sichuan University, Chengdu (Sichuan), China. He has previously worked as Assistant Professor in K. R. Mangalam University (Gurugram, India). He has obtained his doctorate degree from University of Delhi, India, in the field of natural product chemistry. So far, he has published 45 SCI articles with cumulative impact factor of 131. He is also Founder and Chief Editor of *Indo Global Journal of Pharmaceutical Sciences* (UGC-CARE, CNKI, and CrossRef linked Journal) and Review Editor of *Frontiers in Chemistry* and *Frontiers in Oncology*. He has so far reviewed 249 articles for 43 journals. One of his most recent publications is *Analytical Methods for the Assessment of Maillard Reactions in Foods* (Springer).

Chapter 1
Relevance of *Ayurveda*. Therapy of Holistic Application and Classification of Herbs

Abstract The efficacy of herbal extracts has been often questioned due to presence of natural toxin loads like alkaloids and terpenes as well as farm-level applied synthetic crop protection chemicals called pesticides, if any, and fungal toxins like aflatoxins. In addition, the efficacy of synthetically prepared drugs is challenged on account of containing process by-products and exerting side effects. The midway therefore emerges as isolation of active ingredients from herbs or any food source. On the other side, Ayurveda underlines holistic approaches in development of medicine in which organic herbal powders—with or without aqueous herbal extracts—are utilised as drug ingredients. It has been observed that herbs holistically are more effective than the active ingredients isolated from them, if side effects too are considered along with time taken in recuperation. In general, the antimicrobial susceptibility tests do not always confirm the same therapeutic action of drugs (against particular microbial) corresponding to particular disease for which those are traditionally known and holistically underlined in Ayurveda. It might be concluded that the therapeutic system Ayurveda is still quite relevant.

Keywords Active ingredient · Allopathy · Antimicrobial activity · Ayurveda · Kapha · Pitta · Vata

Abbreviations

AI	Active ingredient
AST	Antimicrobial susceptibility test
BHA	Butyl hydroxylanisol
CCRAS	Central Council for Research in Ayurvedic Sciences
GSFA	Codex General Standard for Food Additives
CODEX STAN	Codex standard
COVID-19	COronaVIrus Disease19
DP	Degree of polymerisation
DE	Dextrose equivalent
F&B	Food and beverage

© The Author(s), under exclusive license to Springer Nature Switzerland AG 2021
R. K. Sharma et al., *Indian Herbal Medicines*, Chemistry of Foods,
https://doi.org/10.1007/978-3-030-80918-8_1

FSSAI   Food Safety and Standards Authority of India
MW   Molecular weight
RS   Relative sweetness
SARS-CoV-2   Severe acute respiratory syndrome coronavirus 2
TBHQ   *T*-butyl hydroquinone

1.1   Introduction: Efficacy of Herbs and Synthetic Drugs Facing Challenges

The modern medical science focuses on chemical compounds, called drugs, to be taken individually or in combination to fight against disease. Such compounds are either synthetically prepared or isolated from natural food or herb source.

Drugs are classified as per their specific action or efficacy against ailment. Aspirin and paracetamol are called pain relievers; diazepam and alprazolam are termed as anti-anxiety drugs; diphenylhydantoin and phenobarbital are referred as anti-epileptics; levodopa and amantadine are known as antiparkinsonism; clonidine and guanethidine are recognised as antihypertensive. Besides such a classification of drugs based on their antidisease action, there are also drugs fighting against microbial—bacteria, fungi and viruses—individually known as antibacterial, anti-fungal and antiviral compounds. These compounds are collectively called antibiotics. Despite applying the broad network of instrumental diagnosis and chemotherapeutic cure system, the modern allopath physicians, however, use traditional medicine too in the cases of ailments such as diseases concerning lungs (respiration) heart, liver, nervous system and joint illnesses (pneumonia, hypertension, hepatitis, memory loss and arthritis) (Dorsher and McIntosh 2011). In detail, the following diseases have been reported to be treated by means of allopathic systems, methods and traditional procedures even in recent times with concern to the pandemics by 'COronaVIrus Disease 19' (COVID-19) or 'severe acute respiratory syndrome coronavirus 2' (SARS-CoV-2):

(1) Lung diseases (Ali and Alharbi 2020; Broor et al. 2001; Bussmann and Glenn 2010; Lin et al. 2014; Rigat et al. 2013; Ullman and Frass 2010; Younis et al. 2018; Zhang et al. 2020a b; Wang et al. 2020). In these situations, the concerned treatments include oxygen use and infusion of intravenous fluids with life support. In addition, *Unani* plant-based products and *Ayurvedic* procedures are often reported to show antiviral properties (Kim et al. 2010; Li et al. 2005, 2016). In addition, the Indian Ministry of Ayush has recently communicated that AYUSH 64, a polyherbal formulation developed by the Central Council for Research in Ayurvedic Sciences (CCRAS), has been found to be effective in treatment of mild-to-moderate cases of COVID-19 infection (PIB 2021). The drug, named AYUSH-64 and initially developed for Malaria in 1980, has been reproposed for COVID-19. AYUSH 64 includes *Alstonia scholaris* (aqueous

bark extract), *Picrorhiza kurroa* (aqueous rhizome extract), *Swertia chirata* (aqueous extract of whole plant) and *Caesalpinia crista* (fine-powdered seed pulp)

(2) Heart-related disorders (Arthur et al. 2006; Davidson et al. 2003; de Souza Balbueno et al. 2020; Kim et al. 2010; Kumar et al. 2017; Li et al. 2005; Maron 2015; Mashour et al. 1998; Sharma and Rana 2018; Sharma et al. 2017, 2019). With concern to India, a remarkable part of heart-related diseases— atherosclerosis, coronary artery illness, hypertension and myocardial infarction—are reported to be often studied when speaking of public safety. In this ambit, the promotion of Ayurveda, Unani, Siddha and alternative/traditional therapies is extremely promoted and encouraged (Mahalle et al. 2012)

(3) Liver diseases (Bhatt and Bhatt 1996; Govind 2011; Rajaratnam et al. 2014; Yang et al. 2002; Xiong and Guan 2017)

(4) Illnesses of the human nervous system (Bussmann et al. 2010; Chen et al. 2007; Dorsher and McIntosh 2011; Pandian et al. 2006; Ven Murthy et al. 2010).

The well-known traditional Ayurvedic drugs for treatment of these ailments normally include the major herbs *Emblica officinalis* (amla), *Commiphora mukul* (guggal), *Capparis spinosa* (himsra) with *Cichorium intybus* (chicory), *Centella asiatica* (brahmi) and *Commiphora mukul* with calcined conch (shankh bhasm), respectively. These herbs are also becoming part and parcel of some medicine prescribed by allopath physicians. The reason for adopting Ayurvedic herbal medicine seems to be consideration of side effects, when the patient undergoes prolonged chemotherapy.

There are several cases when overmedication becomes the treatment pathway for a patient. For example, the patient feels partial improvement with intake of tablets, in a few cases of arthritis, particularly spondylitis: anti-inflammatory drug of dichloro-anilino phenyl structure like diclofenac sodium, antibacterial drug of beta-lactum structure like cefixime and analgesic of acetylaminophenol structure like paracetamol. However, after a few months of medication, the patient again undergoes panic situation with feeling of additional complications of swelling in legs and excessive breath filling of lungs with normal body movement, according to unpublished experiences by the first author of this book. Normally, such a disease is diagnosed as increased creatinine and urea level in the blood (blood test report) as well as reduced blood ejection with decreased heart efficiency (Doppler effect test report). The patient is further diagnosed for clogging of arteries and even when angiography report is negative: he or she is subjected to scanning for stress myocardial perfusion, and the medication goes on even when this test report is negative. The physician often prescribes in such a case metoprolol succinate [1-(4-(2-methoxyethyl)phenoxy)-3-(1-methylethyl)amino)-2-propanylsuccinate], furosemide (4-chloro-*N*-furfuryl-5-sulfamoylanthranillic acid) and aspirin (acetyl salicylic acid), in particular if patient feels inconvenience in movement with respiration complication even after the blood pressure is either normal or is made in normal range with regular intake of normal anti-hypertensive like losartan potassium [monopotassium salt of 2-butyl-4 chloro-1-(*p*-(o-1H-tetrazol-5-ylphenyl)benzyl) imidazol-5-methanol] and/or amlodipine besylate [phenyl bisulphite of 2-((2-aminoethoxy) methyl)-4-(2-chlorophenyl)-1,4-dihydro-6-methyl-3,5-pyridinedicarboxylic acid 3-ethyl 5-methyl ester].

Then, after 6-month-prolonged allopathic treatment, in the same condition of painful breathing on body movements, the patient sometimes calls for a *vaidya* (Ayurvedic physician). Then, the disease is diagnosed as *vata* or gastric trouble just by putting fingers on his or her wrist vein. The *vaidya* prescribes *avipathikar churna*—a mixture of sugar 50%, trivrit (*Ipomoea turpethum*) 33.33%, lavang (*Syzygium aromaticum*) 8.33% and each of tejpatra (*Cinnamomum tamala*), elaichi (*Elettaria cardamomum*), vidang (*Embelia ribes*), vid lavan, nagarmoth (*Cyperus rotundus*), amla (*Emblica officinalis*), bibhitak (*Terminalia bellrica*), haritiki (*Terminalia chebula*), pippali (*Piper longum*), kalimirch (*Piper nigrum*) and sunthi (*Zingiber officinale*) 0.75% which makes the patient gradually recuperate within two months, according to unpublished experiences by the first author of this book.

The efficacy of multi-herbal mixtures is often challenged due to presence of natural toxins load like alkaloids and terpenes as well as farm-level applied synthetic crop protection chemicals called pesticides, if any, and fungal toxins due to harsh hot-moist storage environment like aflatoxins. However, allopathic or chemotherapeutic drugs—synthetic compounds as well as bio-product isolates—are too challengeable on account of their side effects in case of a prolonged treatment. Ayurvedic herbal treatment seems to be adoptable when herbs grow in conserved soils of natural dense forests, synthetic pesticide application is avoided, natural pesticide like neem (*Azadirachta indica*) leaf juice is applied on need, and cool-dry environment during storage is maintained. In the case when herbs used are free from synthetic insecticides and have aflatoxins and other naturally occurring contaminants and toxins well in limit (aflatoxin <30 µg/kg; aflatoxin M in milk used for Ayurvedic preparations <0.5 µg/kg; ochratoxin A in wheat, barley and rye if used <20 µg/kg; agaric acid <100 ppm; hydrocyanic acid <5 ppm; hypericine <1 ppm; and safrole <0 ppm) (FSSAI (2011), only then general Ayurvedic herbal preparations might be supposed to act effectively against ailment.

Despite their purity, herbs are often questioned for presence of toxin load of alkaloids, terpenes (mostly skin allergens) and tannins in chemical composition of herbs. This challenge of composition-wise herbal toxins exists in the domain of Ayurvedic practice, how to administer drug dosage.

The efficacy of synthetically prepared allopathic drugs is also challenged on account of containing process by-products and exerting side effects. The midway therefore emerges as isolation of active ingredients from herbs or any food source. This paper is an attempt to observe the consequences of active ingredient (AI) isolation from food source by means of the study of *Charak Samhita* text to understand the holistic classification and application of herbs as well as test reports concerning herbs efficacy. The conceptualisation of observed phenomena is also a declared aim.

1.2 Observation: Limitations of Herbs, Synthetic Drugs and Active Ingredients Isolated

The efficacy of both herbs and synthetic drugs is often challenged due to being loaded with toxins and by-products to a considerable extent. The midway therefore emerges as isolation of active ingredients from herbs or food source.

The natural food additives or active pharmaceutical ingredients isolated from food source turn food processing or drug manufacturing somewhat safe. However, at the same time extraction process turns farm/forest yield unsafe, after active ingredient removal, if presented in market for human consumption. Otherwise, it would be waste, if destroyed. For example, when natural extracts or essential oils are taken out of spices, the de-oiled portion of spices presented in the market for human consumption obviously belongs to inferior quality; otherwise, it will be all waste if destroyed. However, natural isolates can improve processed food or drug quality to an extent if used in place of synthetic additives.

Extraction of digestive enzymes from fruits and vegetables is another example which perhaps explains clearly how the main or almost entire edible part of these food articles is qualitatively deteriorated after enzyme extraction. The papain-free papaya fruit is nowadays normally available in market which cannot meet out the consumer requirement, particularly if it is prescribed by the physician to the patient. Papain or papayotin, the water-soluble and one of the most thermo-stable enzymes with remarkable digestive properties, is obtained as dried and purified latex of papaya. The valuable digestive protein named papain is used not only in pharmaceutical preparations, but also vastly in food industry mainly as meat tenderiser, beer chilling haze (due to presence of proteins) remover and tobacco quality improver. In addition to pharmaceutical and food industries, it is also used in cosmetic, leather and textile industries (Banchhor and Saraf 2008; Dubey et al. 2007; Fernández-Lucas et al. 2017; Li et al. 2016; Mamboya 2012; Manohar et al. 2015; Sangeetha and Abraham 2006; Sim et al. 2000; Xu et al. 2008; Zhang et al. 2020a, b). Therefore, the vast industrial demand for digestive enzyme papain is met out, nowadays, by almost removal of latex from papaya fruit.

The purchase of papaya, which is almost papain-free declared, from market with the idea that it will help in recuperation of a patient suffering particularly from constipation might prove fallacy or delusions.

Extraction of vitamins from food sources to meet out their industrial demand as medicinal, food and feed supplements also might lead to similar situation. For example, green leafy vegetables are remarkable sources of ether-soluble vitamin A and vitamin K (Booth and Suttie 1998; Chandrika et al. 2006; Gupta and Prakash 2009; Kamao et al. 2007; Oboh 2005; Oboh and Akindahunsi 2004; Raju et al. 2007; Schönfeldt and Pretorius 2011; Sim et al. 2020; Suttie 1992; Violi et al. 2016; West and Darnton-Hill 2008). If these vitamins are extracted from green leafy vegetable paste or powders and after extraction, the remaining major part of food article is sold as green vegetable powder (e.g. spinach powder), it is a deceit with consumer who purchases it for recuperation from a disease due to vitamin A and K deficiencies.

Vitamin A and vitamin K deficiencies create abnormalities concerned with night vision and blood clotting, respectively. Any person suffering from these abnormalities is expected to recuperate fast if he regularly consumes green leafy vegetables and their powders. However, vitamin extraction from these food articles might prove this incorrect and unreliable impression. Vitamin C or ascorbic acid present in citrus fruits, tomatoes, potatoes, green leafy vegetables, etc., which can be extracted with water from these natural sources, because it is water-soluble. These food articles, after extraction of vitamin C, are not enough effective for a person suffering from scurvy or any infection requiring ascorbic acid to catalyse cell redox (reduction–oxidation) reaction. It also seems that AI isolated from food source might have their own limitations due to side effects, in absence of various health supportive ingredients left in remaining food after isolation.

In the light of isolation of AI, the 'Go Natural' theme in food sector might also be analysed. The 'Go Natural' theme, through public awakening, opened the vistas for the addition of natural additives instead of synthetic chemicals during processing of food articles. The following situations can be considered in this ambit:

(a)   The natural antioxidant vitamin E (tocopherols) instead of synthetic compound TBHQ (*t*-butyl hydroquinone) and BHA (butyl hydroxylanisol) in oils and fats during frying process (Fig. 1.1)

(b)   Natural flavour enhancer like yeast extracts instead of synthetic compound monosodium glutamate in seasoning process

(c)   Natural flour treatment agent (cysteine) instead of synthetic compound potassium bromate in bread making (Fig. 1.2)

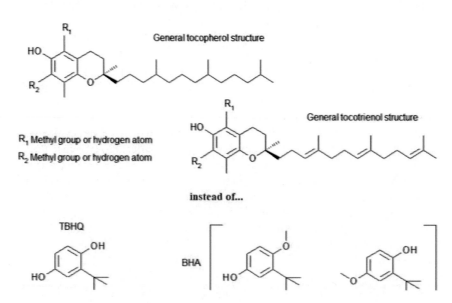

Fig. 1.1 Tocopherols (and also tocotrienols), generally named vitamin E, may be suggested as valid alternatives for TBHQ and BHA in human nutrition

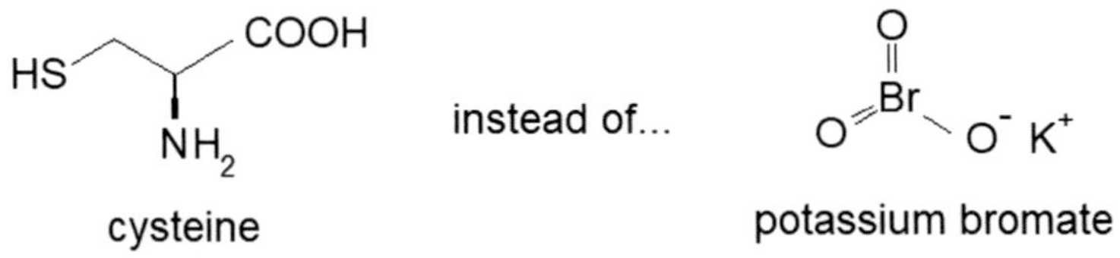

Fig. 1.2 Cysteine can be suggested against potassium bromate in bread making as a natural additive

(d) Natural fortifying vitamins instead of synthetic counterparts in any food article particularly food supplements
(e) Natural colours like annatto or curcumin instead of synthetic colours like tartrazine or sunset yellow.

These examples can demonstrate the possible use of natural processing aids substitution instead of synthetic compounds in food manufacturing, which nowadays are considered as public health concerns. Needless to say, the 'Go Natural' theme contributed a lot to promotion of natural additives application in and removal of synthetic additives from food processing, turning it safe. But there is other side too of this theme. It also contributed to extraction of functional ingredient from farm output, turning it unsafe. It means foods deprived of active ingredients and functionalism are not effective towards fulfilment of consumers' expectations for recuperation if they are ill and advised to eat a functional food—fruit or herb—by physician (normally Ayurvedic physicians).

The foods function in human body as per their ingredients' composition and chemical (minerals, vitamins, etc.) contents of ingredients. In some cases, food functionalism is closely related to a molecule; say papaya's anticonstipation features with enzyme papain, citrus fruits' antiscurvy characteristics with vitamin C and turmeric's antiseptic characteristics with curcumin. The major portion of food after extraction of functional ingredient or specific compound does not function well in the human body. We can say that papain extracted from papaya as tenderiser improves meat quality, but papaya loses efficacy to fight against constipation in human body after considerable extraction of latex, containing papain. Similarly, the major portion of turmeric after considerable extraction of curcumin loses antiseptic nature although curcumin, substituting synthetic yellow colours in processing of food articles, improves quality of processed foods. Although natural food additives turn food processing safe, farm yield is unsafe after their extraction. On the other side, the industry of packaged foods and the pharmaceutical industry have been continuously demanding active isolates, and this industrial demand might pose a serious threat to public health. A patient is perhaps deceived when he or she purchases a functional food on physician's advice, but gets food deprived of functional ingredient.

1.3 Hypothesis: Herbs Holistically More Effective Than Isolated Active Ingredients

In fact, the effects of food are not just for which functional or active ingredient—mineral, vitamin or protein—in how much percentage is present in it. The food—as entire composition of ingredients—acts on the human body. In India, it is said that whey or *chhachh*—the liquid left after extracting out butter from curd (*dahi*)—should be consumed after taking meals. Such health directions are concerned with the routine human actions, not with the components of food articles. Secondly, a little clinically tested risk reduction claims of food articles aim at the presence of a functional additive say ascorbic acid, often extracted from natural products and added to any kind of food ingredients composition including high sugar, salt or fat contents. As such there is also other side of health or nutrition or risk reduction claims that those contribute to extractions of functional ingredients from farm yield, turning it unsafe, for the sake of nominal safety improvement of packaged food articles. This simple food intake phenomenon might be conceived as a therapeutic hypothesis: herbs holistically are more effective than active ingredients isolated from them. It seems that Indian therapeutic system adheres to this principle.

1.4 Studies in the Context of Hypotheses

The six main tongue tastes—sweet, sour, salty, pungent, bitter and astringent—result from several botanical ingredients called herbs and spices, when speaking of cuisine. The most prevalent herbs and spices used along with sugar and/or honey and/or salt in Indian cuisine for taste making are summarised as follows:

(1)   Almond kernels
(2)   Aloe
(3)   Amla
(4)   Asafetida
(5)   Basil leaves
(6)   Bay leaves
(7)   Black pepper
(8)   Caraway
(9)   Cardamom
(10) Chili
(11) Cinnamon
(12) Clove
(13) Coconut
(14) Coriander
(15) Cumin
(16) Dates
(17) Fennel

(18)   Fenugreek
(19)   Garlic
(20)   Ginger
(21)   Lemon juice
(22)   Licorice
(23)   Long pepper
(24)   Mace
(25)   Mustard seeds
(26)   Nutmeg
(27)   Onion
(28)   Pomegranate seeds
(29)   Poppy seeds
(30)   Pumpkin seeds
(31)   Raisins
(32)   Rose petals
(33)   Saffron
(34)   Sesame seeds
(35)   Tamarind
(36)   Thyme
(37)   Turmeric.

It is worth mentioning that Ayurveda, the Indian therapy system which means science of longevity, recognises six tastes: sweet, sour, salty, pungent, bitter and astringent. Dr. David Frawley and Dr. Vasant Lad mention in their book 'The Yoga of Herbs: An Ayurvedic Guide to Herbal Medicines' (Frawley and Lad 1993) that these tastes are considered to be paired in three taste groups on the basis of the common post-digestive effect, *vipaka*. In this ambit, a sweet *vipaka* includes two of the original tastes (sweet and salty); in addition, a pungent *vipaka* is the common attribution for astringent, bitter and pungent tastes, while sour *vipaka* is ascribed to sour taste only (Fig. 1.3). In addition, the following stages (of digestion)—*kapha, pitta* and *vata*—are in strict relation with above-mentioned tastes in terms of alleviation or aggravation of digestion stages.

1.5 Holistic Classification and Application of Herbs in *Charak Samhita*

In accordance to the *Charak Samhita* text, XVI, 43, Frawley and Lad explain the Ayurvedic classification of herbs, as per the observed taste, which is summarised as follows (Frawley and Lad 1993).

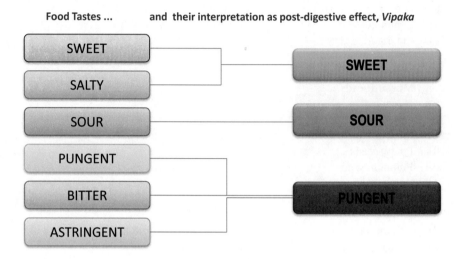

Fig. 1.3 Subdivision of six food tastes by the viewpoint of the post-digestive effects (*vipaka*)

1.5.1   Sweet Taste

1.5.1.1   General Features of Sweet Taste

Sweet taste is nourishing, vitalising, gives contentment, adds bulk to the body and creates firmness. It rebuilds weakness, emaciation and helps those damaged by disease. It is refreshing to the nose, mouth, throat, lips and tongue and relieves fits and fainting. It has been reported to be associated with smell and taste at the same time (Tekiroğlu et al. 2015).

1.5.1.2   Physical and Chemical Features of Sweet Taste

By the physical viewpoint, it may be correlated with specific volumes (Birch et al. 1996). Surely, it has to be correlated with sweet compounds such as sucrose, although many non-carbohydrates are also sweet-tasting molecules (Karl et al. 2020). On the one hand, it has been suggested that water can influence the so-called glycophore/receptor interaction not only in terms of rheology, apparent specific volume and apparent molar volume of tested solutions, but also when speaking of water affinity with specific sugar sites (with possible enhancement of sweetness). On these bases, it has been suggested that glucose has a notable sweetness because it is able to interact directly and/or by means of interposed water molecules with the interested receptor.

On the other side, 'sweet-tasting molecules' are generally carbohydrates. However, not all carbohydrates are sweet-tasting compounds. In general, the detection of sweetness is carried out in the oral cavity having the following targets: simple sugars (mono- and di-saccharides) and complex carbohydrates. A first discrimination should be performed here: simple sugars are generally able to stimulate sweetness receptors with a perceived sweet taste, and these target molecules include glucose, galactose, fructose, maltose, lactose and sucrose. A the same time, it has been reported that short-chain oligosaccharides (degree of polymerisation or DP, $1 < DP \leq 3$, including maltotriose (Pullicin et al. 2019), 4-galactosyl-kojibiose and lactulosucrose (Ruiz-Aceituno et al. 2018) are able to stimulate sweet taste sense on human sensory panels, while the stimulation of sweetness by maltooligosaccharides (DP > 3) is not assured at all. On the contrary, their human panellists report the associated taste as 'starchy'.

The above-discussed situation can explain different evaluations (on the sensorial level at least) when speaking of commercially available maltodextrins, because these mixtures contain different sweet-tasting targets, including approximately 20% in weight of sweet saccharides with $1 < DP \leq 3$. Consequently, the variability in the DP profile should be expected with concern to maltodextrins, in spite of dextrose-equivalent (DE) values. DE evaluation is important enough in this context. DE values concern the degree of acid hydrolysis of maltodextrins, and explicitly the number of reducing ends aldehyde groups relative to pure glucose when concentration is equal. In other words, high DE means high degrees of acid hydrolysis and low average molecular weights (MW), at the same time (Babu et al. 2015; Takeiti et al. 2010). Consequently, it should be expected that high sweetness is perceived if low-MW components are prevailing in the maltodextrin mixture (Griffin and Brooks 1989). After all, molecular weight and chain length appear to be the most relevant factors in determining the sweetness potential of a carbohydrate (Pullicin et al. 2019). As a single example, a disaccharide is perceived to be more sweet than trisaccharides, in turn, exhibited superior sweetness than mixture of oligosaccharide having DP > 3 (Ruiz-Aceituno et al. 2018). However, other physical properties (crystallinity, viscosity) are affected by DE values; for these reasons, DE could be not appropriate when speaking of performance predictions in certain ambits (Takeiti et al. 2010). In fact, different maltodextrins with equal DE values could be differently perceived when speaking of sweetness. More research is surely needed in this ambit, taking also into account the important problem of relative sweetness (RS): the perceived sweetness (by human panellists) of a sweet-tasting molecule or mixture, if compared with RS for sucrose, which is conventionally equal to 100. Because of the influence of several parameters—pH, concentration, serving temperature, matrix effects caused by 'bitter' compounds, stereochemical conformation, concomitant presence of different sweet molecules, etc.—and the subjectivity of human tests, the use of RS scores can be sometimes questionable. On the other hand, RS is mainly used when speaking of simple carbohydrates such as fructose, with the aim of establishing a semi-quantitative scale of sweet compounds and mixed products.

With reference to the 'starchy' perception of short-chain oligosaccharides, an important reflection should concern the role of saliva α-amylase activity for products requiring normal mastication. In fact, it has been reported that starch-related sweet taste appears enhanced for panellists on condition that maltose concentration becomes sensorially higher than in absence of mastication. In other words, starchy-related sweet perception depends on maltose concentration, and this increase can be observed when speaking of chewing gum products, provided that a 2-min mastication is carried out (Aji et al. 2019; Lapis et al. 2014, 2017). In these conditions, saliva α-amylase activity can be positively correlated with maltose concentration and, consequently, with starchy perception. On the other side, the absence of mastication did not affect the perception in terms of starchy sweet taste.

1.5.1.3 Sweetness Perception. From Raw Materials to Final Products

The reliable perception of sweet taste should concern the evaluation of chemical compounds and mixtures, taking into account the subdivision of the whole group of edible sweet-tasting 'commodities' in two categories:

(1) Sweet ingredients, including also sweeteners (actually, these compounds are not explicitly discussed in this review)
(2) Sweet foods and beverages (F&B), also definable 'final products'.

The first category concerns the 'universe' of food ingredients and additives as purchased and used by F&B producers. The second category concerns the final destination of use for sweet ingredients and additive. On these bases, a bijective correspondence between the first category and the groups of F&B possible applications may be displayed (Table 1.1) by means of a common connection: the targeted sweet-tasting molecule (or mixture of sweet-tasting targets). Several papers shown in the selected literature and the 'Codex General Standard for Food Additives' (GSFA, Codex STAN 192-1995, last revision: 2019) have been used for establishing connections between ingredients and final products (1, 35). The main F&B applications have been considered on the basis of Codex standards (CODEX STAN) documents as mentioned in Table 1.1. It should be noted that F&B applications can contemplate the use of more than one single sweet-tasting compound (or related mixture). In addition, several F&B articles are used for subsequent F&B productions as ingredients (maple syrups, toppings, etc.). Consequently, some raw material (ingredient) may be also displayed as final product. Polyols, artificial sweeteners and related mixtures have been excluded for simplification purposes. Several products such as regional recipes, typical desserts, and other F&B which may be defined as expression of cultural heritage such as the South American *dulce de leche* (Garitta et al. 2004) or the Jordanian *labaneh* (Haddad et al. 2021) may be absent in Table 1.1.

Table 1.1 Bijective correspondence between sweet ingredients/additives and final F&B applications, by means of a common connection: the sweet-tasting molecule (sweet-tasting compound(s) of interest) (Codex Alimentarius Commission 1995; Lim and Pullicin 2019)

Ingredient category	Example	Sweet-tasting compound(s) of interest	Main F&B applications (categories)
Refined and raw sugars (Bogdanov et al. 2008; Lee et al. 1970)	Dextrose anhydrous, dried glucose syrup, raw cane sugar, mill white sugar, brown sugar, maple syrup, Demerara sugar	Glucose	Canned Applesauce (CODEX STAN 17-1981) Canned Corned Beef (CODEX STAN 88-1981) Canned Strawberries (CODEX STAN 62-1987) Canned Strawberries (CODEX STAN 62-1987) Canned Tropical Fruit Salad (CODEX STAN 99-1981) Certain Canned Vegetables (CODEX STAN 297-2009): mushrooms Chocolate and Chocolate Products (CODEX STAN 87-1981), as 'sugars' Cocoa Powders (Cocoa) and Dry Mixtures of Cocoa and Sugar (CODEX STAN 105-1981), as 'sugars' Cooked Cured Chopped Meat (CODEX STAN 98-1981) Cooked Cured Pork Shoulder (CODEX STAN 97-1981) Crackers from Marine and Freshwater Fish, Crustaceans and Molluscan Shellfish (CODEX STAN 222-2001), as 'sugar' Cream and Prepared Creams (fermented cream, acidified cream, CODEX STAN 288-1976), as 'sugar' Dairy Fat Spreads (CODEX STAN 253-2006) Edible Fungi and Fungi Products (CODEX STAN 38-1981) Fat Spreads and Blended Spreads (CODEX STAN 256-2007) Fruit Juices and Nectars (CODEX STAN 247-2005) Jams, Jellies and Marmalades (CODEX STAN 296-2009) Luncheon Meat (CODEX STAN 89-1981) Non-fermented Soybean Products (CODEX STAN 32-2015) Pickled Fruits and Vegetables (CODEX STAN 260-2007) Processed Cereal-Based Foods for Infants and Children (CODEX STAN 74-1981) Quick Frozen Strawberries (CODEX STAN 52-1981) Quick Frozen Vegetables (CODEX STAN 320-2015) Soy Protein Products (DEX STAN 175-1989) Special Dietary Foods with Low-Sodium Content, Including salt (CODEX STAN 53-1981), as 'sugars' Sugars (CODEX STAN 212-1999) Vegetable Protein Products (CODEX STAN 174-1989) Wheat Protein Products, Including Wheat Gluten (CODEX STAN 163-1987) Whey Cheeses (CODEX STAN 284-1971)
Honey (Bogdanov et al. 2008; Lee et al. 1970)	Wildflower honey, clover honey	Glucose	Canned Applesauce (CODEX STAN 17-1981) Certain Canned Vegetables (CODEX STAN 297-2009) Cooked Cured Pork Shoulder (CODEX STAN 97-1981) Honey (CODEX STAN 12-1981) Jams, Jellies and Marmalades (CODEX STAN 296-2009) Pickled Fruits and Vegetables (CODEX STAN 260-2007) Processed Cereal-Based Foods for Infants and Children (CODEX STAN 74-1981) Soy Protein Products (DEX STAN 175-1989)

(continued)

Table 1.1 (continued)

Ingredient category	Example	Sweet-tasting compound(s) of interest	Main F&B applications (categories)
Table-top sweeteners (solid, liquid or powdered products)		Glucose	Dairy Fat Spreads (CODEX STAN 253-2006)
Refined and raw sugars (Bogdanov et al. 2008; Lee et al. 1970)	Purified D-fructose	Fructose	Canned Applesauce (CODEX STAN 17-1981) Canned Strawberries (CODEX STAN 62-1987) Dairy Fat Spreads (CODEX STAN 253-2006) Edible Fungi and Fungi Products (CODEX STAN 38-1981) Fat Spreads and Blended Spreads (CODEX STAN 256-2007) Fruit Juices and Nectars (CODEX STAN 247-2005) Jams, Jellies and Marmalades (CODEX STAN 296-2009) Pickled Fruits and Vegetables (CODEX STAN 260-2007) Non-fermented Soybean Products (CODEX STAN 32-2015) Processed Cereal-Based Foods for Infants and Children (CODEX STAN 74-1981) Quick Frozen Strawberries (CODEX STAN 52-1981) Quick Frozen Vegetables (CODEX STAN 320-2015) Soy Protein Products (DEX STAN 175-1989) Special Dietary Foods with Low-Sodium Content, Including salt substitutes (CODEX STAN 53-1981), as 'sugars' Sugars (CODEX STAN 212-1999) Vegetable Protein Products (CODEX STAN 174-1989) Wheat Protein Products, including Wheat Gluten (CODEX STAN 163-1987) Whey Cheeses (CODEX STAN 284-1971)
Invert sugar	Glucose/fructose 1:1 equimolecular mixture	Glucose and fructose	62-1987 Canned Strawberries (CODEX STAN 62-1987) Canned Corned Beef (CODEX STAN 88-1981) Certain Canned Vegetables (CODEX STAN 297-2009): sweet corn; mushrooms Chocolate and Chocolate Products (CODEX STAN 87-1981), as 'sugars' Cocoa Powders (Cocoa) and Dry Mixtures of Cocoa and Sugar (CODEX STAN 105-1981), as 'sugars' Cooked Cured Chopped Meat (CODEX STAN 98-1981) Cooked Cured Pork Shoulder (CODEX STAN 97-1981) Crackers from Marine and Freshwater Fish, Crustaceans and Molluscan Shellfish (CODEX STAN 222-2001), as 'sugar' Cream and Prepared Creams (fermented cream, acidified cream, CODEX STAN 288-1976), as 'sugar' Dairy Fat Spreads (CODEX STAN 253-2006) Fruit Juices and Nectars (CODEX STAN 247-2005) Jams, Jellies and Marmalades (CODEX STAN 296-2009) Luncheon Meat (CODEX STAN 89-1981) Non-fermented Soybean Products (CODEX STAN 32-2015) Soy Protein Products (DEX STAN 175-1989) Sugars (CODEX STAN 212-1999) Vegetable Protein Products (CODEX STAN 174-1989) Wheat Protein Products, Including Wheat Gluten (CODEX STAN 163-1987)

(continued)

Table 1.1 (continued)

Ingredient category	Example	Sweet-tasting compound(s) of interest	Main F&B applications (categories)
Other sweetening sugars (also defined generally 'sweeteners' without specifications) (Park and Yetley 1993; Rumessen 1992)	High fructose corn syrup corn sugar	Various	Bouillon and Consommés (CODEX STAN 117-1981) Canned Tropical Fruit Salad (CODEX STAN 99-1981) Dairy Fat Spreads (CODEX STAN 253-2006) Fruit Juices and Nectars (CODEX STAN 247-2005)
Fermented dairy products (Abrahamson, 2015)	Yogurt	Galactose	Dairy Fat Spreads (CODEX STAN 253-2006) Soy Protein Products (DEX STAN 175-1989) Vegetable Protein Products (CODEX STAN 174-1989) Wheat Protein Products, Including Wheat Gluten (CODEX STAN 163-1987)
Fruits and vegetables	Avocados, sugar beets	Galactose	Dairy Fat Spreads (CODEX STAN 253-2006)
Xanthan gum and other gums (Acosta, and Gross 1995)	Partially hydrolysed Guar gum	Galactose	Follow-Up Formula (CODEX STAN 156-1987)
Refined and raw sugars (Daudé et al. 2012):	White sugar, sugar cane, sugar beet	Sucrose (and related mixtures)	Canned Strawberries (CODEX STAN 62-1987) Blend of Sweetened Condensed Milk and Vegetable Fat (CODEX) 252-2006 Canned Applesauce (CODEX STAN 17-1981) Canned Corned Beef (CODEX STAN 88-1981) Canned Strawberries (CODEX STAN 62-1987) Canned Tropical Fruit Salad (CODEX STAN 99-1981) Certain Canned Vegetables (CODEX STAN 297-2009): mushrooms Chocolate and Chocolate Products (CODEX STAN 87-1981), as 'sugars' Cocoa Powders (Cocoa) and Dry Mixtures of Cocoa and Sugar (CODEX STAN 105-1981), as 'sugars' Cooked Cured Chopped Meat (CODEX STAN 98-1981) Cooked Cured Pork Shoulder (CODEX STAN 97-1981) Crackers from Marine and Freshwater Fish, Crustaceans and Molluscan Shellfish (CODEX STAN 222-2001), as 'sugar' Cream and Prepared Creams (fermented cream, acidified cream, CODEX STAN 288-1976), as 'sugar' Dairy Fat Spreads (CODEX STAN 253-2006) Edible Fungi and Fungi Products (CODEX STAN 38-1981) Fat Spreads and Blended Spreads (CODEX STAN 256-2007) Fruit Juices and Nectars (CODEX STAN 247-2005) Jams, Jellies and Marmalades (CODEX STAN 296-2009) Luncheon Meat (CODEX STAN 89-1981) Non-fermented Soybean Products (CODEX STAN 32-2015) Pickled Fruits and Vegetables (CODEX STAN 260-2007) Processed Cereal-Based Foods for Infants and Children (CODEX STAN 74-1981) Quick Frozen Strawberries (CODEX STAN 52-1981) Quick Frozen Vegetables (CODEX STAN 320-2015) Soy Protein Products (CODEX STAN 175-1989) Special Dietary Foods with Low-Sodium Content, Including salt substitutes (CODEX STAN 53-1981), as 'sugars' Vegetable Protein Products (CODEX STAN 174-1989) Wheat Protein Products, Including Wheat Gluten (CODEX STAN 163-1987) Whey Cheeses (CODEX STAN 284-1971)

(continued)

Table 1.1 (continued)

Ingredient category	Example	Sweet-tasting compound(s) of interest	Main F&B applications (categories)
Different milk types (Goldfein and Slavin 2014)	Cow's milk	Lactose	Blend of Evaporated Skimmed Milk and Vegetable Fat (CODEX STAN 250-2006) Blend of Skimmed Milk and Vegetable Fat in Powdered Form (CODEX STAN 251-2006) Blend of Sweetened Condensed Milk and Vegetable Fat (CODEX) 252-2006 Canned Applesauce (CODEX STAN 17-1981) Canned Corned Beef (CODEX STAN 88-1981) Canned Strawberries (CODEX STAN 62-1987) Cooked Cured Chopped Meat (CODEX STAN 98-1981) Cooked Cured Pork Shoulder (CODEX STAN 97-1981) Dairy Fat Spreads (CODEX STAN 253-2006) Fat Spreads and Blended Spreads (CODEX STAN 256-2007) Infant formula and formula for special dietary purposes intended for infants (formula for special dietary purposes intended for infants, CODEX STAN 72-1981) Jams, Jellies and Marmalades (CODEX STAN 296-2009) Luncheon Meat (CODEX STAN 89-1981) Quick Frozen Strawberries (CODEX STAN 52-1981) Quick Frozen Vegetables (CODEX STAN 320-2015) Soy Protein Products (DEX STAN 175-1989) Sugars (lactose, CODEX STAN 212-1999) Vegetable Protein Products (CODEX STAN 174-1989) Wheat Protein Products, Including Wheat Gluten (CODEX STAN 163-1987) Whey Cheeses (CODEX STAN 284-1971)
Unfermented dairy foods	Mozzarella cheese	Lactose	Dairy Fat Spreads (CODEX STAN 253-2006)
Yeasts and fungi	Dictyostelium mucoroides	Trehalose (surrogate for sucrose)	Dairy Fat Spreads (CODEX STAN 253-2006) Vegetable Protein Products (CODEX STAN 174-1989) Wheat Protein Products, Including Wheat Gluten (CODEX STAN 163-1987)
Mushrooms (Schiraldi et al. 2002)	*Shiitake* (*Lentinula edodes*)	Trehalose	Dairy Fat Spreads (CODEX STAN 253-2006)
Sprouted wheat and barley; malted grains; maltodextrin (BeMille and Whistler 1996)	Malted barley	Maltose	Canned Corned Beef (CODEX STAN 88-1981) Cooked Cured Chopped Meat (CODEX STAN 98-1981) Dairy Fat Spreads (CODEX STAN 253-2006) Luncheon Meat (CODEX STAN 89-1981) Soy Protein Products (DEX STAN 175-1989) Vegetable Protein Products (CODEX STAN 174-1989) Wheat Protein Products, Including Wheat Gluten (CODEX STAN 163-1987)

(continued)

Table 1.1 (continued)

Ingredient category	Example	Sweet-tasting compound(s) of interest	Main F&B applications (categories)
Vegetables/tubers, leaves, roots ()	Sorghum, rice, beans, tapioca, potato	α-Glucans: starch	Canned Baby Foods (CODEX STAN 73-1981) Certain Canned Vegetables (CODEX STAN 297-2009): sweet corn; mushrooms Cheddar (CODEX STAN 263-1966) Cheese (unripened, Including fresh cheese, CODEX STAN 283-1978 and 221-2001) Cooked Cured Chopped Meat (CODEX STAN 98-1981) Cottage Cheese (CODEX STAN 273-1968) Coulommiers (CODEX STAN 274-1969) Cream and Prepared Creams (fermented cream, acidified cream, CODEX STAN 288-1976) Cream Cheese (Rahmfrischkäse, CODEX STAN 275-1973) Dairy Fat Spreads (CODEX STAN 253-2006) Fermented Milks (flavoured, heat treated and non-heat treated, CODEX STAN 243-2003) Infant formula and formula for special dietary purposes intended for infants (formula for special dietary purposes intended for infants, CODEX STAN 72-1981 Instant Noodles (CODEX STAN 249-2006) Luncheon Meat (CODEX STAN 89-1981) Pickled Fruits and Vegetables (CODEX STAN 260-2007) Processed Cereal-Based Foods for Infants and Children (CODEX STAN 74-1981) Vegetable Protein Products (CODEX STAN 174-1989)
Refined and raw sugars	Glucose syrups, corn syrup solids, and maltodextrins)	α-Glucans: hydrolysis products of starch	Dairy Fat Spreads (CODEX STAN 253-2006) Fermented Milks (flavoured, heat treated and non-heat treated, CODEX STAN 243-2003) Sugars (lactose; plantation or white mill sugar, CODEX STAN 212-1999) Vegetable Protein Products (CODEX STAN 174-1989)

Polyols, artificial sweeteners and related mixtures have been excluded for simplification purposes

1.5.1.4 Sweet Taste. Medical-Related Features and Relation with Typical Sweet Herbs

When used in excess, sweet taste creates the following unhealthy states in the human being:

(a) Obesity
(b) Flaccidity
(c) Laziness
(d) Excessive sleep
(e) Heaviness
(f) Loss of appetite
(g) Weak digestion
(h) Abnormal growth of the muscles of mouth and throat
(i) Difficult breathing
(j) Cough
(k) Difficult urination
(l) Intestinal torpor

(m)   Fever due to cold
(n)   Abdominal distention
(o)   Excessive salivation
(p)   Loss of feeling
(q)   Loss of voice
(r)   Goitre
(s)   Swelling of lymph glands
(t)   Legs and neck accumulations in the bladder and blood vessels
(u)   Mucoid accretions in the throat
(v)   Other *kapha*-related diseases.

In the ambit of typical sweet herbs, the following types have to be considered at least:

(1)   Almonds
(2)   Comfrey root
(3)   Dates
(4)   Fennel
(5)   Flaxseed
(6)   Licorice
(7)   Maidenhair fern
(8)   Marshmallow
(9)   Psyllium
(10) Raisins
(11) Sesame seeds
(12) Slippery elm
(13) Solomon's seal.

Sweet taste in herbs can be increased by processing herbs with various forms of raw sugar, honey or cooking them in milk.

1.5.2 Sour Taste

Sour taste improves the taste of food, enkindles the digestive fire, adds bulk to the body, invigorates, awakens the mind, gives firmness to the senses, increases strength, dispels intestinal gas and flatus, gives contentment to the heart, promotes salivation, aids swallowing, moistening and digestion of food, gives nourishment.

Yet when used in excess sour taste makes the teeth sensitive, causes thirst, blinking of the eyes, goose bumps, liquefies *kapha*, aggravates *pitta* and causes a build-up of toxins in the blood. It wastes away the muscles and causes looseness of the body, creates oedema in those weak, injured or in convalescence. It promotes maturation and suppuration of sores, wounds, burns, fractures and other injuries. It causes a burning sensation in the throat, chest and heart. Sour taste occurs largely from the presence of various acids in plants, like acid fruits.

Typical sour herbs include:

(a) Hawthorn berries
(b) Lemon
(c) Lime
(d) Raspberries
(e) Rose hips.

Sour taste in herbs can be increased by preparing herbs in fermentation as herbal wines or as tincture in alcohol (whose taste is sour).

1.5.3 Salty Taste

Salty taste promotes digestion, is moistening and enkindles digestive fire. It works as a sedative, laxative and de-obstruent. Salty taste alleviates *vata*, relieves stiffness, contractions, softens accumulations and nullifies all other taste. It promotes salivation, liquefies *kapha*, cleanses the vessels, softens all the organs of the body and gives taste to food.

Yet when used in excess it aggravates *pitta*, causes stagnation of blood and creates thirst, fainting and the sensation of burning, erosion and wasting of muscles. It aggravates infectious skin conditions, causes symptoms of poisoning, causes tumours to break open, makes the teeth fall, decreases virility, obstructs the functioning of the senses, causes wrinkling of the skin, greying and falling of the hair. Salty taste promotes bleeding diseases, hyperacidity of digestion, inflammatory skin diseases, gout and other mainly *pitta* diseases. Salty taste is a mineral taste; it is very rare in plants as a primary taste.

Typical salty substances include:

(a) Epsom salt
(b) Irish moss
(c) Kelp
(d) Rock salt
(e) Sea salt
(f) Seaweed.

Salty taste in herbs can be increased by adding salt to herbal preparations.

1.5.4 Pungent Taste

Pungent taste is cleansing to the mouth, enkindles digestive fire, purifies food, promotes nasal secretions, causes tears and gives clarity to the senses. It helps cure diseases of intestinal torpor, obesity, abdominal swelling and excessive liquid in the body. It helps discharge oily, sweaty and sticky waste products. It gives taste to food, stops itching, helps the resolution of skin growths, kills worms, is germicidal, moves blood clots and blood stagnation, breaks up obstructions, opens the vessels and alleviates *kapha*.

Yet when used in excess causes a weakening of virility by its post digestive effect. It causes delusion, weariness, languor, emaciation. Pungent taste causes fainting, prostration, loss of consciousness and dizziness. It burns the throat, generates a burning sensation in the body, diminishes strength and causes thirst. It creates various burning sensations, tremors and piercing and stabbing pains throughout the body. Pungent taste arises from various aromatic oils. They often become spices and condiments. Pungent tastes include acrid, spicy and aromatic.

Typical pungent herbs include:

(1)   Angelica
(2)   Asafetida
(3)   Basil
(4)   Bayberry
(5)   Bay leaves
(6)   Black pepper
(7)   Camphor
(8)   Cardamom
(9)   Cayenne
(10)   Cinnamon
(11)   Cloves
(12)   Coriander
(13)   Cumin
(14)   Ephedra
(15)   Eucalyptus
(16)   Garlic
(17)   Ginger
(18)   Horseradish
(19)   Mustard
(20)   Onion
(21)   Oregano
(22)   Peppermint
(23)   Prickly ash
(24)   Rosemary sage
(25)   Sassafras
(26)   Spearmint
(27)   Thyme
(28)   Valerian.

1.5.5 Bitter Taste

Sweet taste restores the sense of taste. It is detoxifying, antibacterial, germicidal and kills worms. It relieves fainting, burning sensation, itch, inflammatory skin conditions and thirst. Bitter taste creates tightness of the skin and muscles. It is antipyretic, febrifuge; it enkindles digestive fire, promotes digestion of toxins, purifies lactation, helps scrap away fat and remove toxic accumulations in fat, marrow, lymph, sweat, urine, excrement, *pitta* and *kapha*.

Yet when used in excess it causes a wasting away of all the tissue elements of the body. Bitter taste produces roughness in the vessels, takes away strength, causes emaciation, weariness, delusion, and dizziness, dryness of the mouth and other diseases of *vata*.

Typical bitter herbs include:

(a) Aloe
(b) Barberry
(c) Blessed thistle
(d) Blue flag
(e) Chaparral
(f) Chrysanthemum
(g) Dandelion
(h) Echinacea
(i) Gentian
(j) Golden seal
(k) Pao d'arco
(l) Peruvian bark
(m) Rhubarb
(n) Rue
(o) Tansy
(p) White poplar
(q) Yarrow
(r) Yellow dock.

1.5.6 Astringent Taste

Astringent taste is a sedative, stops diarrhoea, aids in healing of joints, promotes the closing and healing of sores and wounds. It is drying, firming, contracting. It alleviates *kapha*, *pitta* and stops bleeding. Astringent taste promotes absorption of bodily fluids.

Yet when used in excess, it causes drying of the mouth produces pain in the heart, causes constipation, weakens the voice, obstructs channels of circulation, makes the skin dark, weakens vitality and causes premature ageing. Astringent taste causes the retention of gas, urine, faces and creates emaciation, weariness, thirst and stiffness. It causes *Vata*-diseases like paralysis, spasms and convulsions.

Typical astringent herbs include:

(a)   Cranesbill
(b)   Lotus seeds
(c)   Mullein
(d)   Plantain
(e)   Pomegranate
(f)   Raspberry leaves
(g)   Sumach
(h)   Uva ursi
(i)   White pond lily
(j)   White oat bark
(k)   Witch hazel.

1.5.7   Combined Tastes

There are also combined tastes which could be summarised as follows:

(a)   Sweet and pungent tastes sometimes combine as with cinnamon, fennel, ginger and onion. Such herbs are particularly good for *vata*.
(b)   Sweet and astringent often combine, as with comfrey, lotus, slippery elm and white pond lily. Such herbs are particularly good for *pitta* but may be hard to digest.
(c)   Sweet and bitter sometimes combine, as with licorice. These herbs are particularly good for *pitta*.
(d)   Sweet and sour combine in various fruits like hawthorn and oranges. They are good for *vata*.
(e)   Pungent and bitter sometimes combine as with motherwort, mugwort, wormwood and yarrow. Such herbs have a strong effect on *kapha*.
(f)   Pungent and astringent combine occasionally, as with bayberry and sage. They work on *kapha*.
(g)   Bitter and astringent often combine, as in many diuretics; such herbs include golden seal and uva ursi. They work mainly on *pitta*.
(h)   Some herbs possess three or more tastes. Herb of multiple tastes often possesses powerful healing action like garlic.

Frawley and Lad explain how, according to Ayurveda taste making, herbs taken alone or with food, with precaution, act as drugs (Frawley and Lad 1993). Those might be used to cure diseases, but their excessive consumption can lead to side effects in body. Even it may happen with regular consumption of a fruit or its juice. For example, pomegranate juice regularly consumed for blood-related ailments might cause constipation. The use of herbs, like drugs, varies according to situation. The fenugreek seed decoction is good for digestive, respiratory, urinary and reproductive systems, but it may cause abortion and promote vaginal bleeding if consumed by pregnant women. In this way, the classification and application of drug in Ayurveda are holistic in nature and concern with three post-digestive effects in human body.

1.6 Efficacy of Extracts of Herbs in the Light of Antimicrobial Susceptibility Tests

There is a fundamental difference between the concepts of health in Ayurveda and allopathy. The existence of health, according to Ayurveda, is the outcome of balance (state of being well in range) between three post-digestive effects or doshas called *vata*, *pitta* and *kapha*.

The movement balance in body is considered due to *vata*. If it is deranged, body feels cold and suffers from disease like flatulence and rheumatism.

The energy or heat balance in body is considered due to *pitta*. If it is deranged, metabolism in body gets so affected that body feels heat or burning sensation giving rise to diseases concerned with liver, heart, eyes and skin.

The water balance in body is considered due to *kapha*. In case it is deranged, ear, nose, throat and lungs are affected (Frawley and Lad 1993). The vaidya (Ayurvedic physician) diagnoses post-digestive effects as *vata*, *pitta* or *kapha* by putting his or her fingers on patent's wrist vein and prescribes medicine accordingly. On the other hand, according to allopathic concept, health exists due to control of pathogenic germs or microbial (bacteria, fungi and viruses). Allopath doctors many times prescribe medicine after instrumental diagnosis of individual microbial.

The antimicrobial susceptibility tests (ASTs) to determine efficacy of herbal extract against microbial have been conducted by so many researchers. These research reports might be used to study any correlation between Ayurvedic and allopathic concepts of health as per following viewpoints:

(i)   Whether antimicrobial activity of an Ayurvedic formulation is average of antimicrobial activities of individual constituent herbs

(ii)   Whether a renowned Ayurvedic formulation leads to increase in antimicrobial activity compared to any formulation of lesser number of constituent herbs (or in other words whether a generally prescribed formulation surely leads to increase in antimicrobial activity).

According to test reports on in vitro antimicrobial analysis of *triphala*, the formulation consisting of Amla (*Emblica officinalis*), Harad (*Terminalia chebula*) and Bibhitak (*Terminalia bellerica*), and comparison with its individual constituents authored by Darshana Mahajan and Sapna Jain can be summarised as follows (Mahajan and Jain 2015):

(a)   The antimicrobial activity of *triphala* against *Escherichia coli*, *Pseudomonas aeruginosa*, *Klebsiella pneumoniae* and fungi such as *Candida albicans* is much higher than the average of antimicrobial activities of individual herbal constituent.

(b)   The above-mentioned antimicrobial action is almost equal to average of constituent antimicrobial activities against *Staphylococcus aureus* and *Bacillus subtilis* and the *Aspergillus niger* fungus. That means antimicrobial activity of a formulation in some pathogen cases may appreciably vary from the average of antimicrobial activities of constituent herbs.

According to test report on in vitro antimicrobial activity of *mahasudarshan churna* (a renowned formulation of 13 herbs—*Curcuma longa, Berberis aristata, Solanum xanthocarpum, Solanum indicum, Zingiber officinale, Piper nigrum, Piper longum, Fagonia Arabica, Picrorhiza Kurroa* and *Holarrhena antidysenterica* added to *triphala*) and comparison with *triphala*, the average of antimicrobial activity of *mahasudarshan churna* against ten common pathogens (*Escherichia coli, Staphylococcus aureus, Enterobacter aerogenes, Pseudomonas aeruginosa, Bacillus subtilis, Klebsiella pneumoniae, Salmonella typhi, Staphylococcus epidermidis, Salmonella typhimurium* and *Proteus vulgaris*) is almost 65% of that of *triphala* alone (Tambekar and Dahikar 2011). That means a generally prescribed formulation does not always lead to increase in antimicrobial activity.

1.7   Discussion and Conclusion: *Ayurveda* is Relevant

In the context of herbs' drug-like properties, herbs and spices obviously cannot be regarded safe unless these are safe as per food requirements. First of all, these materials are organic in nature or free from synthetic insecticide residue; secondly, these herbs and spices grown and stored in proper temperature-moisture condition or almost free from crop contaminants (aflatoxins, patulin, ochratoxin). Additionally, they contain naturally occurring toxic substances (agaric acid, hydrocyanic acid, hypericine, safrole, etc.) quite below the maximum permissible limits (Sharma and Parisi 2017). The AI isolates as allopathic drugs too are obviously unsafe because those contain the principle functional ingredient but not the entire range of natural ingredients responsible for safety of a natural herb (Dudeja et al. 2016). For example, the synthetic and nature-identical vanilla flavours contain vanillin, the chief flavour ingredient in natural vanilla. However, in absence of the entire range of health supportive ingredients, the alone principle functional flavour ingredient vanillin, mostly used in ice-creams, can exert carcinogenic effect if consumed regularly. Similarly, AI isolates for pharmaceutical purposes might exert side effects in human body on regular dosing.

Besides these practical aspects for hypothesising—herbs holistically are more effective than active ingredients isolated from them—reported studies on holistic classification and application of herbs in the *Charak Samhita* text, and test reports on efficacy of herbal extracts provide theoretical strength to this hypothesis. The holistic classification of herbs—primarily based on three post-digestive effects and six tastes—covers the cure aspects of diseases concerned with several malfunctions of organs. However, the diagnosis of post-digestive effects *vata*, *pitta* and *kapha* is not instrumental in nature, and therefore, it requires immense human skill. Despite this shortcoming, Ayurveda is ever-relevant because renowned Ayurvedic formulations do not essentially lead to increase antimicrobial activity.

In fact, antimicrobial activity is lowered in the cases of some of the renowned Ayurvedic formulations like *Mahasudarshan churna*. Perhaps it may be the reason for slow action and low side effect exertion of Ayurvedic drugs on patient's body. *Mahasudarshan churna* formulation case related to decrease in average antimicrobial activity against ten microbial is different than *triphala* formulation case related to increase in antimicrobial activity against some of the microbial, namely *E. coli*, *P. aeruginosa*, *K. pneumoniae* and *C. ablicans*. The efficacy of *triphala*, traditionally known for improvement of digestion system and solution of constipation, is evident in the light of its antimicrobial activity against digestive tract affecting microbial. It might be conclusively said that antimicrobial activity enhancement is possible by herbal formulations; therefore, extraction of active ingredients from herbs is not essential for this purpose. In other words, the Ayurvedic cure is not based on targeting the pathogen agent(s). It might be perhaps a serious subject in the field of medicine: how microbials, so-called pathogens, enter in body in the state of imbalance of *vipaka* or post-digestive effects *vata*, *pitta* and *kapha*.

References

Abrahamson A (2015) Galactose in dairy products. Dissertation. Swedish University of Agricultural Sciences, Faculty of Natural Resources and Agricultural Sciences, Department of Food Science, Uppsala, Sweden. Available http://stud.epsilon.slu.se/. Accessed 25 Feb 2021

Acosta PB, Gross KC (1995) Hidden sources of galactose in the environment. Eur J Pediatr 154(2):S87–S92. https://doi.org/10.1007/BF02143811

Aji GK, Warren FJ, Roura E (2019) Salivary α-amylase activity and starch-related sweet taste perception in humans. Chem Sens 44(4):249–256. https://doi.org/10.1093/chemse/bjz010

Ali I, Alharbi OM (2020) COVID-19: disease, management, treatment, and social impact. Sci Total Environ 728:138861. https://doi.org/10.1016/j.scitotenv.2020.138861

Arthur HM, Patterson C, Stone JA (2006) The role of complementary and alternative therapies in cardiac rehabilitation: a systematic evaluation. Eur J Prev Cardiol 13(1):3–9. https://doi.org/10.1097/01.hjr.0000198917.67987.f8

Babu AS, Parimalavalli R, Jagannadham K, Rao JS (2015) Chemical and structural properties of sweet potato starch treated with organic and inorganic acid. J Food Sci Technol 52(9):5745–5753. https://doi.org/10.1007/s13197-014-1650-x

Banchhor M, Saraf S (2008) Potentiality of papain as an antiaging agent in cosmetic formulation. Pharmacogn Rev 2(4):266–270

BeMiller JN, Whistler RL (1996) Carbohydrates. In: Fennema R (ed) Food chemistry. Marcel Dekker, New York, pp 157–224

Bhatt AD, Bhatt NS (1996) Indigenous drugs and liver disease. Ind J Gastroenterol 15(2):63–67

Birch GG, Parke S, Siertsema R, Westwell JM (1996) Specific volumes and sweet taste. Food Chem 56(3):223–230. https://doi.org/10.1016/0308-8146(96)00018-0

Bogdanov S, Jurendic T, Sieber R, Gallmann P (2008) Honey for nutrition and health: a review. J Am Coll Nutr 27:677–689. https://doi.org/10.1080/07315724.2008.10719745

Booth SL, Suttie JW (1998) Dietary intake and adequacy of vitamin K. J Nutr 128(5):785–788. https://doi.org/10.1093/jn/128.5.785

Broor S, Pandey RM, Ghosh M, Maitreyi RS, Lodha R, Singhal T, Kabra SK (2001) Risk factors for severe acute lower respiratory tract infection in under-five children. Ind Pediatr 38(12):1361–1369

Bussmann R, Glenn A (2010) Medicinal plants used in Peru for the treatment of respiratory disorders. Rev Peruv Biol 17(3):331–346

Bussmann RW, Glenn A, Sharon D (2010) Healing the body and soul: traditional remedies for magical ailments, nervous system and psychosomatic disorders in Northern Peru. Afr J Pharm Pharmacol 4(9):580–629

Chandrika UG, Svanberg U, Jansz ER (2006) In vitro accessibility of β-carotene from cooked Sri Lankan green leafy vegetables and their estimated contribution to vitamin A requirement. J Sci Food Agric 86(1):54–61. https://doi.org/10.1002/jsfa.2307

Chen FP, Chen TJ, Kung YY, Chen YC, Chou LF, Chen FJ, Hwang SJ (2007) Use frequency of traditional Chinese medicine in Taiwan. BMC Health Serv Res 7(1):1–11. https://doi.org/10.1186/1472-6963-7-26

Codex Alimentarius Commission (1995) Codex general standard for food additives, codex STAN 192-1995, last revision: 2019. Codex Alimentarius Commission, Food and Agriculture Organization of the United Nations, Rome, and the World Health Organization, Geneva. Available http://www.fao.org/gsfaonline/docs/CXS_192e.pdf. Accessed 25th Feb 2021

Daudé D, Remaud-Siméon M, André I (2012) Sucrose analogs: an attractive (bio) source for glycodiversification. Nat Prod Rep 29(9):945–960. https://doi.org/10.1039/C2NP20054F

Davidson P, Hancock K, Leung D, Ang E, Chang E, Thompson DR, Daly J (2003) Traditional Chinese medicine and heart disease: what does Western medicine and nursing science know about it? Eur J Cardiovasc Nurs 2(3):171–181. https://doi.org/10.1016/S1474-5151(03)00057-4

de Souza Balbueno MC, Junior KDCP, de Paula CC (2020) Evaluation of the efficacy of Crataegus oxyacantha in dogs with early-stage heart failure. Homeopath 109(04):224–229. https://doi.org/10.1055/s-0040-1710021

Dorsher PT, McIntosh PM (2011) Acupuncture's effects in treating the sequelae of acute and chronic spinal cord injuries: a review of allopathic and traditional Chinese medicine literature. Evid Based Compl Altern Med 2011:ID 428108. https://doi.org/10.1093/ecam/nep010

Dubey VK, Pande M, Singh BK, Jagannadham MV (2007) Papain-like proteases: applications of their inhibitors. Afr J Biotechnol 6(9):1077–1086

Dudeja P, Gupta RK, Minhas AS (eds) (2016) Food safety in the 21st century: public health perspective. Academic Press, London, San Diego, Cambridge, Kidlington

Fernández-Lucas J, Castañeda D, Hormigo D (2017) New trends for a classical enzyme: Papain, a biotechnological success story in the food industry. Trends Food Sci Technol 68:91–101. https://doi.org/10.1016/j.tifs.2017.08.017

Frawley D, Lad V (1993) The yoga of herbs. Lotus Press, Wilmot

FSSAI (2011) Food safety and standards (contaminants, toxins, residues) regulation 2011, 2.2. Ministry of Health and Family Welfare, Food Safety and Standards Authority of India (FSSAI), New Delhi

Garitta L, Hough G, Sánchez R (2004) Sensory shelf life of dulce de leche. J Dairy Sci 87(6):1601–1607. https://doi.org/10.3168/jds.S0022-0302(04)73314-7

Goldfein KR, Slavin JL (2014) Why sugar is added to food: food science 101. Compr Rev Food Sci Food Saf 14(5):644–656. https://doi.org/10.1111/1541-4337.12151

Govind P (2011) Medicinal plants against liver diseases. Int Res J Pharm 2(5):115–121

Griffin VK, Brooks JR (1989) Production and size distribution of rice maitodextrins hydrolyzed from milled rice flour using heat-stable alpha-amylase. J Food Sci 54(1):190–193. https://doi.org/10.1111/j.1365-2621.1989.tb08599.x

Gupta S, Prakash J (2009) Studies on Indian green leafy vegetables for their antioxidant activity. Plant Food Hum Nutr 64(1):39–45. https://doi.org/10.1007/s11130-008-0096-6

Haddad MA, Yamani MI, Jaradat DMM, Obeidat M, Abu-Romman SM, Parisi S (2021) Jordan dairy products and traceability. Labaneh, a concentrated strained Yogurt. In: Food traceability in Jordan—current perspectives, Springer International Publishing, Cham, Switzerland. https://doi.org/10.1007/978-3-030-66820-4_3

Kamao M, Suhara Y, Tsugawa N, Uwano M, Yamaguchi N, Uenishi K, Ishida H, Sasaki S, Okano T (2007) Vitamin K content of foods and dietary vitamin K intake in Japanese young women. J Nutr Sci Vitaminol 53(6):464–470. https://doi.org/10.3177/jnsv.53.464

Karl CM, Wendelin M, Lutsch D, Schleining G, Dürrschmid K, Ley JP, Krammer GE, Lieder B (2020) Structure-dependent effects of sweet and sweet taste affecting compounds on their sensorial properties. Food Chem X 7:100100. https://doi.org/10.1016/j.fochx.2020.100100

Kim HY, Eo EY, Park H, Kim YC, Park S, Shin HJ, Kim K (2010) Medicinal herbal extracts of Sophorae radix, Acanthopanacis cortex, Sanguisorbae radix and Torilis fructus inhibit coronavirus replication in vitro. Antivir Ther 15(5):697–709. https://doi.org/10.3851/IMP1615

Kumar V, Kshemada K, Ajith KG, Binil RS, Deora N, Sanjay G, Jaleel A, Muraleedharan TS, Anandan EM, Mony RS, Valiathan MS, Santhosh KTR, Kartha CC (2017) Amalaki rasayana, a traditional Indian drug enhances cardiac mitochondrial and contractile functions and improves cardiac function in rats with hypertrophy. Sci Rep 7(1):1–17. https://doi.org/10.1038/s41598-017-09225-x

Lapis TJ, Penner MH, Lim J (2014) Evidence that humans can taste glucose polymers. Chem Sens 39(9):737–747. https://doi.org/10.1093/chemse/bju031

Lapis TJ, Penner MH, Balto AS, Lim J (2017) Oral digestion and perception of starch: effects of cooking, tasting time, and salivary α-amylase activity. Chem Sens 42(8):635–645. https://doi.org/10.1093/chemse/bjx042

Lee CY, Shallenberger RS, Vittum MT (1970) Free sugars in fruits and vegetables. N Y Food Life Sci Bull 1:1–12

Li SY, Chen C, Zhang HQ, Guo HY, Wang H, Wang L, Wang L, Zhang X, Hua SN, Yu J, Xiao PG, Li RS, Tan X (2005) Identification of natural compounds with antiviral activities against SARS-associated coronavirus. Antivir Res 67(1):18–23. https://doi.org/10.1016/j.antiviral.2005.02.007

Li Y, Yu J, Goktepe I, Ahmedna M (2016) The potential of papain and alcalase enzymes and process optimizations to reduce allergenic gliadins in wheat flour. Food Chem 196:1338–1345. https://doi.org/10.1016/j.foodchem.2015.10.089

Lim J, Pullicin AJ (2019) Oral carbohydrate sensing: beyond sweet taste. Physiol Behav 202:14–25. https://doi.org/10.1016/j.physbeh.2019.01.021

Lin LT, Hsu WC, Lin CC (2014) Antiviral natural products and herbal medicines. J Tradit Compl Med 4(1):24–35. https://doi.org/10.4103/2225-4110.124335

Mahajan D, Jain S (2015) Antimicrobial analysis of triphala and comparison with its individual constituents. Ind J Pharm Biolog Res 3(3):55–60. https://doi.org/10.30750/ijpbr.3.3.9

Mahalle NP, Kulkarni MV, Pendse NM, Naik SS (2012) Association of constitutional type of Ayurveda with cardiovascular risk factors, inflammatory markers and insulin resistance. J Ayurveda Integr Med 3(3):150–157. https://doi.org/10.4103/0975-9476.100186

Mamboya EAF (2012) Papain, a plant enzyme of biological importance: a review. Am J Biochem Biotechnol 8(2):99–104. https://doi.org/10.3844/ajbbsp.2012.99.104

Manohar CM, Prabhawathi V, Sivakumar PM, Doble M (2015) Design of a papain immobilized antimicrobial food package with curcumin as a crosslinker. PLoS ONE 10(4):e0121665. https://doi.org/10.1371/journal.pone.0121665

Maron BJ (2015) Importance and feasibility of creating hypertrophic cardiomyopathy centers in developing countries: the experience in India. Am J Cardiol 116(2):332–334. https://doi.org/10.1016/j.amjcard.2015.04.027

Mashour NH, Lin GI, Frishman WH (1998) Herbal medicine for the treatment of cardiovascular disease: clinical considerations. Arch Intern Med 158(20):2225–2234. https://doi.org/10.1001/archinte.158.20.2225

Murthy MRV, Ranjekar PK, Ramassamy C, Deshpande M (2010) Scientific basis for the use of Indian ayurvedic medicinal plants in the treatment of neurodegenerative disorders: 1. Ashwagandha. Cent Nerv Sys Agents Med Chem (Formerly Curr Med Chem Cent Nerv Sys Agents) 10(3):238–246.

Oboh G (2005) Effect of blanching on the antioxidant properties of some tropical green leafy vegetables. LWT-Food Sci Technol 38(5):513–517. https://doi.org/10.1016/j.lwt.2004.07.007

Oboh G, Akindahunsi AA (2004) Change in the ascorbic acid, total phenol and antioxidant activity of sun-dried commonly consumed green leafy vegetables in Nigeria. Nutr Health 18(1):29–36. https://doi.org/10.1177/026010600401800103

Pandian JD, Bose S, Daniel V, Singh Y, Abraham AP (2006) Nerve injuries following intramuscular injections: a clinical and neurophysiological study from Northwest India. J Peripher Nerv Sys 11(2):165–171. https://doi.org/10.1111/j.1085-9489.2006.00082.x

Park YK, Yetley EA (1993) Intakes and food sources of fructose in the United States. Am J Clin Nutr 58(5):737S-747S. https://doi.org/10.1093/ajcn/58.5.737S

PIB (2021) Ministry of Ayush responds to Frequently Asked Questions (FAQs) about "Ayush-64". The polyherbal drug AYUSH 64 has been found to be useful in treating mild to moderate cases of Covid 19 in clinical trials. Press Information Bureau (PIB), Government of India, New Delhi. Available https://pib.gov.in/PressReleaseIframePage.aspx?PRID=1715849. Accessed 18 May 2021

Pullicin AJ, Penner MH, Lim J (2019) The sweet taste of acarbose and maltotriose: relative detection and underlying mechanism. Chem Sens 44(2):123–128. https://doi.org/10.1093/chemse/bjy081

Rajaratnam M, Prystupa A, Lachowska-Kotowska P, Zaluska W, Filip R (2014) Herbal medicine for treatment and prevention of liver diseases. J PreClin Clin Res 8(2):55–60. https://doi.org/10.5604/18982395.1135650

Raju M, Varakumar S, Lakshminarayana R, Krishnakantha TP, Baskaran V (2007) Carotenoid composition and vitamin A activity of medicinally important green leafy vegetables. Food Chem 101(4):1598–1605. https://doi.org/10.1016/j.foodchem.2006.04.015

Rigat M, Vallès J, Iglésias J, Garnatje T (2013) Traditional and alternative natural therapeutic products used in the treatment of respiratory tract infectious diseases in the eastern Catalan Pyrenees (Iberian Peninsula). J Ethnopharmacol 148(2):411–422. https://doi.org/10.1016/j.jep.2013.04.022

Ruiz-Aceituno L, Hernandez-Hernandez O, Kolida S, Moreno FJ, Methven L (2018) Sweetness and sensory properties of commercial and novel oligosaccharides of prebiotic potential. LWT 97:476–482. https://doi.org/10.1016/j.lwt.2018.07.038

Rumessen JJ (1992) Fructose and related food carbohydrates: sources, intake, absorption, and clinical implications. Scand J Gastroenterol 27:819–828. https://doi.org/10.3109/00365529209000148

Sangeetha K, Abraham TE (2006) Chemical modification of papain for use in alkaline medium. J Mol Catal B Enzym 38(3–6):171–177. https://doi.org/10.1016/j.molcatb.2006.01.003

Schiraldi C, Di Lernia I, De Rosa M (2002) Trehalose production: exploiting novel approaches. Trends Biotechnol 20(10):420–425. https://doi.org/10.1016/S0167-7799(02)02041-3

Schönfeldt HC, Pretorius B (2011) The nutrient content of five traditional South African dark green leafy vegetables—a preliminary study. J Food Comp Anal 24(8):1141–1146. https://doi.org/10.1016/j.jfca.2011.04.004

Sharma RK, Parisi S (2017) Botanical ingredients and herbs in India. Foods or drugs? In: Toxins and contaminants in Indian food products. Springer International Publishing, Cham, pp 25–34

Sharma RK, Rana BK (2018) Studies on antimicrobial activity and kinetics of inhibition by plant products in India (1990–2016). J AOAC Int 101(4):948–955. https://doi.org/10.5740/jaoacint.17-0449

Sharma M, Kartha CC, Mukhopadhyay B, Goyal RK, Gupta SK, Ganguly NK, Dhalla NS (2017) India's march to halt the emerging cardiovascular epidemic. Circ Res 121:913–916. https://doi.org/10.1161/CIRCRESAHA.117.310904

Sharma RK, Micali M, Pellerito A, Santangelo A, Natalello S, Tulumello R, Singla RK (2019) Studies on the determination of antioxidant activity and phenolic content of plant products in India (2000–2017). J AOAC Int 102(5):1407–1413. https://doi.org/10.1093/jaoac/102.5.1407

Sim YC, Lee SG, Lee DC, Kang BY, Park KM, Lee JY, Kim MS, Chanh IS, Rhee JS (2000) Stabilization of papain and lysozyme for application to cosmetic products. Biotechnol Lett 22(2):137–140. https://doi.org/10.1023/A:1005670323912

Sim M, Lewis JR, Prince RL, Levinger I, Brennan-Speranza TC, Palmer C, Bondonno CP, Bondonno NP, Devine A, Ward NC, Byrnes E, Schultz CS, Woodman R, Croft K, Hodgson JM, Blekkenhorst LC (2020) The effects of vitamin K-rich green leafy vegetables on bone metabolism: a 4-week randomised controlled trial in middle-aged and older individuals. Bone Rep 12:100274. https://doi.org/10.1016/j.bonr.2020.100274

Suttie JW (1992) Vitamin K and human nutrition. J Am Diet Assoc 92(5):585–590

Takeiti CY, Kieckbusch TG, Collares-Queiroz FP (2010) Int J Food Prop 13(2):411–425. https://doi.org/10.1080/10942910802181024

Tambekar DM, Dahikar SB (2011) Antibacterial activity of some Indian Ayurvedic preparations against enteric bacterial pathogens. J Adv Pharm Technol Res 2(1):24–29. https://doi.org/10.4103/2231-4040.79801

Tekiroğlu SS, Özbal G, Strapparava C (2015) Exploring sensorial features for metaphor identification. In: Proceedings of the third workshop on Metaphor in NLP, Denver, Colorado, 5 June 2015, pp 31–39. Available https://www.aclweb.org/anthology/W15-1400/. Accessed 05 May 2021

Ullman D, Frass M (2010) A review of homeopathic research in the treatment of respiratory allergies. Altern Med Rev 15(1):48–58

Violi F, Lip GY, Pignatelli P, Pastori D (2016) Interaction between dietary vitamin K intake and anticoagulation by vitamin K antagonists: is it really true?: a systematic review. Medicine 95(10):e2895. https://doi.org/10.1097/MD.0000000000002895

Wang Z, Chen X, Lu Y, Chen F, Zhang W (2020) Clinical characteristics and therapeutic procedure for four cases with 2019 novel coronavirus pneumonia receiving combined Chinese and Western medicine treatment. Biosci Trends. https://doi.org/10.5582/bst.2020.01030

West KP, Darnton-Hill I (2008) Vitamin A deficiency. In: Semba RD., Bloem MW, Piot P (eds) Nutrition and health in developing countries. Nutr Health Series, Humana Press. https://doi.org/10.1007/978-1-59745-464-3_13

Xiong F, Guan YS (2017) Cautiously using natural medicine to treat liver problems. World J Gastroenterol 23(19):3388–3395. https://doi.org/10.3748/wjg.v23.i19.3388

Xu W, Bai W, Guo F, Luo Y, Yuan Y, Huang K (2008) A papaya-specific gene, papain, used as an endogenous reference gene in qualitative and real-time quantitative PCR detection of transgenic papayas. Eur Food Res Technol 228(2):301–309. https://doi.org/10.1007/s00217-008-0935-6

Yang ZC, Yang SH, Yang SS, Chen DS (2002) A hospital-based study on the use of alternative medicine in patients with chronic liver and gastrointestinal diseases. Am J Chin Med 30(04):637–643. https://doi.org/10.1142/S0192415X02000569

Younis W, Asif H, Sharif A, Riaz H, Bukhari IA, Assiri AM (2018) Traditional medicinal plants used for respiratory disorders in Pakistan: a review of the ethno-medicinal and pharmacological evidence. Chin Med 13(1):1–29. https://doi.org/10.1186/s13020-018-0204-y

Zhang ZJ, Wu WY, Hou JJ, Zhang LL, Li FF, Gao L, Wu XD, Shi JY, Zhang R, Long HL, Lei M, Wu WY, Guo DA, Chen KX, Hofmann LA, Ci ZH (2020) Active constituents and mechanisms of Respiratory Detox Shot, a traditional Chinese medicine prescription, for COVID-19 control and prevention: network-molecular docking-LC–MSE analysis. J Integr Med 18(3):229–241. https://doi.org/10.1016/j.joim.2020.03.004

Zhang Z, Bai G, Xu D, Cao Y (2020) Effects of ultrasound on the kinetics and thermodynamics properties of papain entrapped in modified gelatin. Food Hydrocoll 105:105757. https://doi.org/10.1016/j.foodhyd.2020.105757

Chapter 2
Indian Herbal Extracts as Antimicrobial Agents

Abstract This chapter discusses several features concerning the most interesting compounds with claimed and demonstrable antimicrobial action, provided these substances are of vegetable origin. This discussion concerns different types of molecules and materials, and some differences may be also evaluated when speaking of ambits of use. The available uses concern life and health extension, protection of crops, enhancement of durability values concerning food products, etc. Anyway, each antimicrobial power or feature has to be reliable on the basis of dedicated clinical experiments and in vivo or in vitro studies. The most powerful additive or antimicrobial contrasting agent cannot completely eradicate microbial survival and related effects, similarly to the inevitability of food perishability, in accordance with the Parisi's First Law of Food Degradation. With relation to herbal extracts, antimicrobial properties are generally ascribed to more than ten classes, from the chemical viewpoint, including alkaloids, coumarins, phenolics, polyamines, tannins and terpenes. On the international ground, research on herbal extracts with some antimicrobial claim has been reported interesting results and reliable data. By the viewpoint of Indian researchers, several facts have been reported when speaking of *Acacia nilotica*, *Datura stramonium*, *Azadirachta indica* extracts, and two mixtures—*triphala* and *mahasudarshan churna*. Traditional herbal medicines and remedies in India are able to exhibit good antimicrobial properties. Recent efforts by Indian researchers still continue in this ambit.

Keywords Alkaloid · Antibiotic susceptibility test · Antimicrobial activity · *Ayurveda* · Herbal extract · Inhibition · Terpenes

Abbreviations

AST	Antimicrobial Susceptibility Test
AOAC	Association of Official Analytical Chemists
CLSI	Clinical and Laboratory Standards Institute
COVID-19	COronaVIrus Disease 19
ISO	International Organization for Standardization

© The Author(s), under exclusive license to Springer Nature Switzerland AG 2021
R. K. Sharma et al., *Indian Herbal Medicines*, Chemistry of Foods,
https://doi.org/10.1007/978-3-030-80918-8_2

MIC   Minimum inhibitory concentration
SARS-CoV-2   Severe acute respiratory syndrome coronavirus 2
TLC   Thin-layer chromatography

2.1   Introduction to Herbal Extracts and Antimicrobial Properties

The most interesting compounds with claimed and demonstrable antimicrobial features concern different types of molecules and materials, and some differences may be also evaluated when speaking of ambits of use (medical structures and therapies; agricultural practices; food and pharma processing plants; etc.). The available uses concern life and health extension, protection of crops, enhancement of durability values concerning food products, etc. Anyway, each antimicrobial power or feature has to be reliable on the basis of dedicated clinical experiments and in vivo or in vitro studies. In fact, the most powerful additive or antimicrobial contrasting agent cannot completely eradicate microbial survival and related effects, similarly to the inevitability of food perishability, in accordance with the Parisi's First Law of Food Degradation (Anonymous 2021; Baglio 2018; Pellerito et al. 2019; Srivastava 2019).

With reference to India, herbal extracts and related mixtures have been often considered with concern to claimed antimicrobial properties and related kinetics of inhibition. In general, the following plants have been reported to have similar features, even if it has to be considered that the below-mentioned list is not exhaustive, because of the possibility of different plant mixtures (Amalraj and Gopi 2017; Balammal et al. 2012; Banerjee and Nigam 1978; Bele et al. 2009; Chaturvedi et al. 2011; Dabur et al. 2007; Ghoshal et al. 1996; Gul and Bakht 2015; Gupta et al. 2015; Jain et al. 2015; Khatri and Juvekar 2016; Lalitha 2012; Mahadevan and Sridhar 1982; Malik and Ahmed 2016; Parekh and Chanda 2007; Rana et al. 1997; Rani and Khullar 2004; Sharma and Rana 2018; Sharma et al. 2009, 2013, 2019; Sohni et al. 1995; Sushmitha et al. 2013; Tirupathi et al. 2011; Joshi et al. 2011):

(a)   *Curcuma longa*, especially with reference to extractable essential oils
(b)   *Piper longum* fruits
(c)   *Aegle marmelos* leaves, especially with reference to extractable essential oils
(d)   *Azadirachta indica* (neem), taking into account that its role as antimicrobial medicine could sometimes be shared with other plants in mixture
(e)   *Acacia nilotica*
(f)   *Aloe vera*
g)   *Annona squamosa*
h)   *Caesalpinia pulcherrima*
(i)   *Datura stramonium*
(j)   *Eugenia caryophyllata*

(k)   *Holarrhena antidysenterica*
(l)   *Myristica fragrans*
(m)   *Punica granatum*
(n)   *Justicia zelanica*
(o)   *Lantana camara*
(p)   *Piper betel*
(q)   *Salmalia mala barica*
(r)   *Saraca asoca*
s)   *Terminalia arjuna.*

Basically, reliable antimicrobial effects mean and concern reliable/demonstrable susceptibility of life forms, parasites and also viruses if placed against one or more associated active principles (Sharma and Rana 2018; Sharma et al. 2019). However, it has to be clarified that each antimicrobial effect is substantially the inhibition of a specific agent, while the eradication is not among the aims of antimicrobial susceptibility tests. This concept should be taken into account, especially at present when the world is facing pandemics by 'COronaVIrus Disease 19' (COVID-19) or 'severe acute respiratory syndrome coronavirus 2' (SARS-CoV-2) by means of chemical therapies and also allopathic systems and traditional procedures. In particular, many diseases of human interest have to be considered in this moment, and the collateral effect of selected microbial agents has to be critically considered:

(1)   Lung diseases (Younis et al. 2018). Current treatments include oxygen use and infusion of intravenous fluids with life support, although the use of *Unani* plant-based products and *Ayurvedic* procedures is often considered.
(2)   Heart-related diseases—atherosclerosis, coronary artery illness, hypertension and myocardial infarction (Mashour et al. 1998). The Indian perspective is extremely interesting in the ambit of public safety: the promotion of Ayurveda, Unani, Siddha, and alternative/traditional therapies is promoted and encouraged (Mahalle et al. 2012).
(3)   Illnesses of the human nervous system (Chen et al. 2007).
(4)   Liver diseases (Xiong and Guan 2017).

Antimicrobial susceptibility is studied by means of officially recognised protocols such as those created by the International Organization for Standardization (ISO), the Clinical and Laboratory Standards Institute (CLSI) and the Association of Official Analytical Chemists (AOAC) Official Methods of Analysis. In general, considered methods are based on the concept of diffusion of the antimicrobial agent(s) on a selected ground. Consequently, some interesting procedures involve the use of paper chromatography and thin-layer chromatography (TLC). Substantially, the antimicrobial substance is placed into a container with the aim of allowing it to flow onto a paper or thin-layer surface with the target life form. Should an approximately circular are of inhibition (IZ) appear on the support, the targeted microorganism would be really inhibited and the inhibition diameter of IZ area is a quantitative expression of microbial inhibition (Sharma and Rana 2018; Sharma et al. 2019). This test, named antibiotic susceptibility test (AST), can have some problem when speaking of differences between in vitro and in vivo tests; it was initially considered for antibiotics in blood systems, and it can be also used for the evaluation of amino acids, vitamins and so on (Lalitha 2012).

Actually, AST methods are generally defined as dilution or diffusion procedures (Bariş et al. 2006; Bauer et al. 1966; Freixa et al. 1996; Garrod and Heatley 1944; Goodall and Len 1946; Grover and Moore 1962; Harborne 1973; Lalitha 2012; Lourens et al. 2004; Mishra and Tiwari 1992; Müller and Hinton 1941; Nair et al. 1991; Nene and Thapilyal 2002; Norrel and Messley 1997; Nostro et al. 2000; Rao and Suryalakshmi 1998; Rios et al. 1998; Salie et al. 1996; Tenover et al. 1995; Verma and Kharwar 2006):

(1) Diffusion methods: agar dilution, broth micro-dilution and broth macro-dilution techniques
(2) Dilution methods: agar disc diffusion, agar well diffusion, poison food technique and bio-autography.

Dilution methods are recommended when speaking of accurate determination of antimicrobial activity. This measure, obtainable by means of different AST methods (Lalitha 2012; Lalitha et al. 1997), is named minimum inhibitory concentration (MIC): the lowest amount of a claimed antimicrobial substance which should inhibits microbial growth (minimum expected reduction: 80%) at 37 °C for 24 h, with relative humidity: 100% (Sharma and Rana 2018; Sharma et al. 2019), under proper test controls (say, incubation for 24 h at 37 °C and with relative humidity equal to 100%).

2.2 Active Principles from Vegetable Organisms Against Microbial Growth

With relation to herbal extracts, antimicrobial properties are generally ascribed to more than ten classes, from the chemical viewpoint (Sharma and Rana 2018; Sharma et al. 2019):

- Alkaloids
- Coumarins
- Flavones, flavonoids and flavonols
- Glycosides
- Isothiocyanates
- Phenols and phenolic acids
- Polyamines
- Saponins
- Sterols
- Tannins
- Terpenes
- Thiosulfinates.

Chemical extraction from selected vegetable parts (tubers, rhizomes, roots …) should be performed by means of low-boiling point organic solvents: dichloromethane; methanol in mixture with water; ethanol; hexane; acetone in water; chloroform (Dilika et al. 1997; Eloff 1998a; Härmälä et al. 1992; Parekh et al. 2006; Serkedijeva and Manolova 1992). Naturally, the use of one or another solution depends on the polarity or the absence of polarity of targeted compounds, on thermolability, the possibility of successive extraction with increasing polarity of solvents and on the possible bio-damage caused by the use of one or another solvent (Chang et al. 2002; Dabur et al. 2004; Das et al. 2010; de Paiva et al. 2004; Eloff 1998a, b; Green 2004; Hammer et al. 1999; Kaur and Kapoor 2002; Parekh et al. 2005; Raman 2006; Sofowora 1993; Trease and Evans 1989).

2.3   Herbal Extracts and Antimicrobial Properties. The Indian Perspective

The international research on herbal extracts with some antimicrobial claim has been reported interesting results and reliable data (Witkowska-Banaszczak et al. 2005). By the viewpoint of Indian researchers (Fig. 2.1), the following facts have been revealed and evaluated (Chaturvedi et al. 2011; Dabur et al. 2007; Mahajan and Jain 2015; Mishra et al. 2016; Sharma and Rana 2018; Sharma et al. 2013, 2019; Shete and Chitanand 2014; Tambekar and Dahilar 2011):

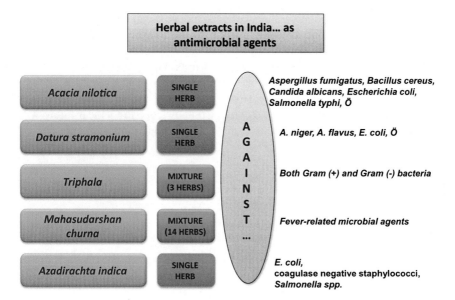

Fig. 2.1 Herbal extract in India with reference to potential action against microbial agents responsible for human diseases. The list of microbial agents may be not completely exhaustive

(a) *Acacia nilotica* extracts are apparently good antimicrobial agents if considered against the following microorganisms: *Aspergillus fumigatus, A. Niger, A. flavus, Bacillus cereus, Candida albicans, Escherichia coli, Klebsiella aerogenes, Proteus vulgaris, Pseudomonas aeruginosa, Salmonella typhi, Shigella boydii,* and *Staphylococcus aureus.*

(b) The following life forms—*A. niger, A. flavus, E. coli, Fusarium culmorum, S. aureus, P. aeruginosa,* and *Rhizopus stolonifer*—can be also inhibited by *Datura stramonium* extracts.

(c) The so-called *triphala*—an herbal mixture of three different fruits: *Emblica officinalis; Terminalia belerica;* and *T. chebula*—is claimed and apparently able to successfully inhibit gram-positive and gram-negative bacteria. It should be also noted that separated herbal extracts can be good antimicrobial inhibitors. In addition, *triphala* appears to be a good antioxidant mixture.

(d) Another herbal composition in India—*Mahasudarshan Churna,* obtained as mixture of 14 different vegetable species—is reported to be good against fever, probably demonstrating that the traditional Indian medicine (named Ayurvedic tradition) does not correspond to allopathic systems.

(e) *Azadirachta indica* is reported to be a good inhibitor agent against *E. coli,* coagulase negative *staphylococci* and *Salmonella* spp.

2.4 Herbal Extracts and Antimicrobial Properties. Conclusions

Based on all above-mentioned references and studies, it can be inferred that traditional herbal medicines and remedies in India are able to exhibit good antimicrobial properties, even if some differences have been noted between negligible or unsure MIC inhibition of certain Indian extracts in AST tests and claimed chemotherapeutic effects. The Ayurvedic approach is quite different (Sharma and Rana 2018; Sharma et al. 2019) from currently allopathy. However, recent efforts by Indian researchers still continue with possible news and correlated knowledge in the near future with concern to the use of traditional herbs against different microbial diseases.

References

Amalraj A, Gopi S (2017) Medicinal properties of *Terminalia arjuna* (Roxb.) Wight & Arn.: a review. J Trad Comp Med 7(1):65–78. https://doi.org/10.1016/j.jtcme.2016.02.003

Anonymous (2021) Parisi's First law of food degradation valuable to establish adequate controls concerning food durability. Inside Lab Manag 21(1):17

Baglio E (2018) Honey: processing techniques and treatments. In: Chemistry and technology of honey production. Springer International Publishing, Cham

Balammal G, Babu MS, Reddy PJ (2012) Analysis of herbal medicines by modern chromatographic techniques. Int J Preclin Pharmaceut Res 3(1):50–63

Banerjee A, Nigam SS (1978) Antimicrobial efficacy of the essential oil of *Curcuma longa*. Ind J Med Res 68:864–866

Bariş Ö, Güllüce M, Şahin F, Özer H, Kiliç H, Özkan H, Sökmen M, Özbek T (2006) Biological activities of the essential oil and methanol extract of *Achillea biebersteinii* Afan. (Asteraceae). Turk J Biol 30(2):65–73

Bauer AW, Kirby WMM, Sheries JC, Turck M (1966) antibiotic susceptibility testing by a standardized single disk method. Am J Clin Pathol 45:493–496

Bele AA, Jadhav VM, Nikam SR, Kadam VJ (2009) Antibacterial potential of herbal formulation. Res J Microbiol 4(4):164–167. https://doi.org/10.3923/jm.2009.164.167

Chang CC, Yang MH, Wen HM, Chern JC (2002) Estimation of total flavonoid content in propolis by complementary colorimetric methods. J Food Drug Anal 10:178–182

Chaturvedi P, Bag A, Rawat V, Jyala NS, Satyavali V, Jha PK (2011) Antibacterial effects of *Azadirachta indica* leaf and bark extracts in clinical isolates of diabetic patients. Natl J Integr Res Med 2(1):5–9

Chen FP, Chen TJ, Kung YY, Chen YC, Chou LF, Chen FJ, Hwang SJ (2007) Use frequency of traditional Chinese medicine in Taiwan. BMC Health Serv Res 7(1):1–11. https://doi.org/10.1186/1472-6963-7-26

Dabur R, Ali M, Singh H, Gupta J, Sharma GL (2004) A novel antifungal pyrrole derivative from *Datura metel* leaves. Pharmazie 59(7):568–570

Dabur R, Gupta A, Mandal TK, Singh DD, Bajpai V, Gurav AM, Lavekar GS (2007) Antimicrobial activity of some Ind medicinal plants. Afr J Trad Comp Altern Med 4(3):313–318. https://doi.org/10.4314/ajtcam.v4i3.31225

Das K, Tiwari RKS, Srivastava DK (2010) Techniques for evaluation of medicinal plant products as antimicrobial agent: current methods and future trends. J Med Plants Res 4:104–111. https://doi.org/10.5897/JMPR09.030

De Paiva SR, Lima LA, Figueiredo MR, Kaplan MAC (2004) Plumbagin quantification in roots of *Plumbago scandens* L. obtained by different extraction techniques. Ann Acad Bras Cienc 76(3):499–504. https://doi.org/10.1590/S0001-37652004000300004

Dilika F, Afolayan AJ, Meyer JJM (1997) Comparative antibacterial activity of two Helichrysum species used in male circumcision in South Africa. S Afr J Bot 63(3):158–159. https://doi.org/10.1016/S0254-6299(15)30728-6

Eloff JN (1998a) A sensitive and quick microplate method to determine the minimal inhibitory concentration of plant extracts for bacteria. Plant Med 64:711–713. https://doi.org/10.1055/s-2006-957563

Eloff JN (1998b) Which extractant should be used for the screening and isolation of antimicrobial components from plants? J Ethnopharmacol 60(1):1–8. https://doi.org/10.1016/S0378-8741(97)00123-2

Freixa B, Vila R, Vargas L, Lozano N, Adzet T, Caniguera S (1996) Screening for antifungal activity of nineteen Latin American plants. Phytother Res 12(6):427–430. https://doi.org/10.1002/(SICI)1099-1573(199809)12:6%3c427::AID-PTR338%3e3.0.CO;2-X4

Garrod LP, Heatley NG (1944) Bacteriological methods in connexion with penicillin treatment. Brit J Surg 32(125):117–124. https://doi.org/10.1002/bjs.18003212527

Ghoshal S, Krishna Prasad PN, Lakshmi V (1996) Antiamoebic activity of Piper longum fruits against Entamoeba histolytica in vitro and in vivo. J Enthopharmacol 50(3):167–170. https://doi.org/10.1016/0378-8741(96)01382-7

Goodall RR, Len AA (1946) A microchromatographic method for the detection and approximate determination of the different penicillins in a mixture. Nature 158(4019):675–676. https://doi.org/10.1038/158675a0

Green RJ (2004) Antioxidant activity of peanut plant tissues. Masters Thesis, North Carolina State University, Raleigh, USA

Grover RK, Moore JD (1962) Toximetric studies of fungicides against brown rot organism *Sclerotina fruticola*. Phytopathol 52:876–880

Gul P, Bakht J (2015) Antimicrobial activity of turmeric extract and its potential use in food industry. J Food Sci Technol 52(4):2272–2279. https://doi.org/10.1007/s13197-013-1195-4

Gupta A, Mahajan S, Sharma R (2015) Evaluation of antimicrobial activity of Curcuma longa rhizome extract against Staphylococcus aureus. Biotechnol Rep 6:51–55. https://doi.org/10.1016/j.btre.2015.02.001

Hammer KA, Carson CF, Riley TV (1999) Antimicrobial activity of essential oils and other plant extracts. J Appl Microbiol 86(6):985–990. https://doi.org/10.1046/j.1365-2672.1999.00780.x

Harborne JB (1973) Phytochemical methods. Chapman & Hall, London, UK, and New York, USA, pp 49–188. https://doi.org/10.1007/978-94-009-5921-7

Härmälä P, Vuorela H, Törnquist K, Hiltunen R (1992) Choice of solvent in the extraction of Angelica archangelica roots with reference to calcium blocking activity. Plant Med 58(2):76–183. https://doi.org/10.1055/s-2006-961424

Jain I, Jain P, Bisht D, Sharma A, Srivastava B, Gupta N (2015) Use of traditional Ind plants in the inhibition of caries-causing bacteria-Streptococcus mutans. Braz Dent J 26(2):110–115. https://doi.org/10.1590/0103-6440201300102

Joshi B, Sah GP, Basnet BB, Bhatt MR, Sharma D, Subedi K, Pandey J, Malla R (2011) Phytochemical extraction and antimicrobial properties of different medicinal plants: Ocimum sanctum (Tulsi), Eugenia caryophyllata (Clove), Achyranthes bidentata (Datiwan) and Azadirachta indica (Neem). J Microbiol Antimicrob 3:1–7

Kaur C, Kapoor HC (2002) Anti-oxidant activity and total phenolic content of some Asian vegetables. Int J Food Sci Technol 37(2):153–161. https://doi.org/10.1046/j.1365-2621.2002.00552.x

Khatri DK, Juvekar AR (2016) Kinetics of inhibition of monoamine oxidase using curcumin and ellagic acid. Pharmacogn Mag 12(2):S116–S120. https://doi.org/10.4103/0973-1296.182168

Lalitha MK (2012) Manual on Antimicrobial Susceptibility Testing. Department of Microbiology, Christian Medical College Vellore,Tamil Nadu, pp 10–13

Lalitha MK, Manayani DJ, Priya L, Jesudason MV, Thomask S (1997) E test as an alternative to conventional MIC determination for surveillance of drug resistant Streptococcus pneumoniae. Ind J Med Res 106:500–503

Lourens ACU, Reddy D, Başer KHC, Vijoen AM, Van Vuuren SF (2004) In vitro biological activity and essential oil composition of four indigenous South African Helichrysum species. J Ethnopharmacol 95(2–3):253–258. https://doi.org/10.1016/j.jep.2004.07.027

Mahadevan A, Sridhar R (1982) Methods in physiological plant pathology, 2nd edn. Sivakami Publications, Madras

Mahajan D, Jain S (2015) Antimicrobial Analysis of Triphala and comparison with its individual constituents. Ind J Pharm Biolog Res 3(3):55–60. https://doi.org/10.30750/ijpbr.3.3.9

Mahalle NP, Kulkarni MV, Pendse NM, Naik SS (2012) Association of constitutional type of Ayurveda with cardiovascular risk factors, inflammatory markers and insulin resistance. J Ayurveda Integr Med 3(3):150–157. https://doi.org/10.4103/0975-9476.100186

Malik N, Ahmed S (2016) Antimicrobial activity of Carica papaya, Piper nigrum and Datura stramonium plants on drug resistant pathogens isolated from clinical specimens. IOSR J Biotechnol Biochem 2(6 II):1–6

Mashour NH, Lin GI, Frishman WH (1998) Herbal medicine for the treatment of cardiovascular disease: clinical considerations. Archiv Int Med 158(20):2225–2234. https://doi.org/10.1001/archinte.158.20.2225

Mishra M, Tiwari SN (1992) Toxicity of Polyalthia longifolia against fungal pathogens of rice. Ind Phytopath 45:56–61

Mishra S, Anuradha J, Tripathi S, Kumar S (2016) In vitro antioxidant and antimicrobial efficacy of Triphala constituents: Emblica officinalis, Terminalia belerica and Terminalia chebula. J Pharmacogn Phytochem 5(6):273–277

Müller HJ, Hinton J (1941) A protein-free medium for primary isolation of the Gonococcus and Meningococcus. Proc Soc Exp Biol Med 48(1):330–333

Nair MG, Safir GR, Siqueira JO (1991) Isolation and identification of vesicular-arbuscular mycorrhiza-stimulatory compounds from clover (*Trifolium repens*) roots. Appl Environ Microb 57(2):434–439

Nene Y, Thapilyal L (2002) Poisoned food technique of fungicides in plant disease control, 3rd edn. Oxford & IBH Publishing Company, New Delhi

Norrel SA, Messley KE (1997) Microbiology laboratory manual: principles and applications. Prentice Hall, Upper Saddle River

Nostro A, Germano MP, D'Angelo V, Marino A, Cannatelli MA (2000) Extraction methods and bioautography for evaluation of medicinal plant antimicrobial activity. Lett Appl Microbiol 30(1):379–384. https://doi.org/10.1046/j.1472-765x.2000.00731.x

Parekh J, Chanda SV (2007) In vitro antimicrobial activity and phytochemical analysis of some Ind medicinal plants. Turk J Biol 31(1):53–58

Parekh J, Jadeja D, Chanda S (2005) Efficacy of aqueous and methanol extracts of some medicinal plants for potential antibacterial activity. Turk J Biol 29(4):203–210

Parekh J, Karathia N, Chanda S (2006) Screening of some traditionally used medicinal plants for potentially anti bacterial activity. Ind J Pharm Sci 68:832–834. https://doi.org/10.4103/0250-474X.31031

Pellerito A, Dounz-Weigt R, Micali M (2019) Food sharing: chemical evaluation of durable foods. Springer International Publishing, Cham

Raman N (2006) Phytochemical methods. New Indian Publishing Agencies, New Delhi, p 19

Rana BK, Singh UP, Taneja V (1997) Antifungal activity and kinetics of inhibition by essential oil isolated from leaves of Aegle marmelos. J Ethnopharmacol 57(1):29–34. https://doi.org/10.1016/S0378-8741(97)00044-5

Rani P, Khullar N (2004) Antimicrobial evaluation of some medicinal plants for their anti-enteric potential against multi-drug resistant Salmonella typhi. Phytotherapy Research: An International Journal Devoted to Pharmacological and Toxicological Evaluation of Natural Product Derivatives. Phytother Res 18(8):670–673. https://doi.org/10.1002/ptr.1522

Rao AVVS, Suryalakshmi A (1998) A textbook of biochemistry: for medical students, Eight. UBS Publisher's Distributors Pvt. Ltd., New Delhi

Rios JL, Recio MC, Villar A (1998) Screening methods for natural products with antimicrobial activity: a review of the literature. J Ethnopharmacol 23(2–3):127–149. https://doi.org/10.1016/0378-8741(88)90001-3

Salie F, Eagles PFK, Lens HMJ (1996) Preliminary antimicrobial screening of four South African Asteraceae species. J Ethnopharmacol 52(1):27–33. https://doi.org/10.1016/0378-8741(96)01381-5

Serkedijeva J, Manolova N (1992) Anti-influenza virus effect of some propolis constituents and their analogues (esters of substituted cinnamic acids). J Nat Prod 55(3):294–297. https://doi.org/10.1021/np50081a003

Sharma D, Lavania AA, Sharma A (2009) In vitro comparative screening of antibacterial and antifungal activities of some common plants and weeds extracts. Asian J Exp Sci 23:169–172. https://doi.org/10.1055/s-0033-1336478

Sharma RA, Sharma P, Yadav A (2013) Antimicrobial screening of sequential extracts of *Datura stramonium* L. Int J Pharm Pharm Sci 5(2):401–404

Sharma RK, Micali M, Pellerito A, Santangelo A, Natalello S, Tulumello R, Singla RK (2019) Studies on the determination of antioxidant activity and phenolic content of plant products in India (2000–2017). J AOAC Int 102(5):1407–1413. https://doi.org/10.1093/jaoac/102.5.1407

Sharma RK, Rana BK (2018) Studies on antimicrobial activity and kinetics of inhibition by plant products in India (1990–2016). J AOAC Int 101(4):948–955. https://doi.org/10.5740/jaoacint.17-0449

Shete HG, Chitanand MP (2014) Antimicrobial activity of some commonly used Indian Spices. Int J Curr Microbiol Appl Sci 3:765–770

Sofowora A (1993) Medicinal plants and traditional medicine in Africa. Spectrum Books Ltd., lbadan, p 289

Sohni YR, Kaimal P, Bhatt RM (1995) The antiamoebic effect of a crude drug formulation of herbal extracts against *Entamoeba histolytica* in vitro and in vivo. J Enthopharmacol 45(1):43–52. https://doi.org/10.1016/0378-8741(94)01194-5

Srivastava PK (2019) Status report on bee keeping and honey processing. Development Institute, Ministry of Micro, Small & Medium Enterprises, Government of India 107, Industrial Estate, Kalpi Road, Kanpur-208012, p 64. Available http://msmedikanpur.gov.in/cmdatahien/reports/diffIndustries/Status%20Report%20on%20Bee%20keeping%20&%20Honey%20Processing%202019-2020.pdf. Accessed 10 May 2021

Sushmita S, Vidyamol KK, Ranganayaki P, Vijayragavam R (2013) Phytochemical extraction and antimicrobial properties of Azadirachta indica (Neem). Glob J Pharmacol 7(3):316–320

Tambekar DM, Dahikar SB (2011) Antibacterial activity of some Ind Ayurvedic preparations against enteric bacterial pathogens. J Adv Pharm Technol Res 2(1):24–29. https://doi.org/10.4103/2231-4040.79801

Tenover FC, Swenson JM, O'Hara CM, Stocker SA (1995) Ability of commercial and reference antimicrobial susceptibility testing methods to detect vancomycin resistance in enterococci. J Clin Microb 33(66):1524–1527

Tirupathi RG, Suresh BK, Kumar JU, Sujana P, Rao AV, Sreedhar AS (2011) Anti–microbial principles of selected remedial plants from Southern India. Asian Pac J Trop Biomed 1(4):298–305. https://doi.org/10.1016/S2221-1691(11)60047-6

Trease GE, Evans WC (1989) Pharmacognosy, 13th edn. Bailliere Tindall, London, pp 176–180

Verma VC, Kharwar RN (2006) Efficacy of neem leaf extract against it's own fungal endophyte Curvularia lunata. J Agric Technol 2(2):329–335

Witkowska-Banaszczak E, Bylka W, Matławska I, Goślińska O, Muszyński Z (2005) Antimicrobial activity of Viola tricolor herb. Fitoter 76(5):458–461. https://doi.org/10.1016/j.fitote.2005.03.005

Xiong F, Guan YS (2017) Cautiously using natural medicine to treat liver problems. World J Gastroenterol 23(19):3388–3395. https://doi.org/10.3748/wjg.v23.i19.3388

Younis W, Asif H, Sharif A, Riaz H, Bukhari IA, Assiri AM (2018) Traditional medicinal plants used for respiratory disorders in Pakistan: a review of the ethno-medicinal and pharmacological evidence. Chin Med 13(1):1–29. https://doi.org/10.1186/s13020-018-0204-y

Chapter 3
Indian Herbal Extract as Antioxidant Agents

Abstract This chapter discusses the importance of phenolic substances in vegetable products in India with reference to claimed antioxidant power. The most interesting compounds with claimed and demonstrable antioxidant power concern different molecules in the ambit of natural materials, especially when speaking of phenolics compounds. Antioxidant power is an interesting argument in the ambit of food and pharma industries, with important implications in the field of cosmetic applications. Antioxidant substances aim at the diminution of damages suffered by organic substrates caused by oxygen attack and consequent oxidative processes. In other terms, each substance able to delay oxidation has to be considered as an antioxidant agent. The importance of such a substance is correlated with the decrease of acute and chronic effects of oxidation on organic surfaces. The remarkable interest of antioxidant compounds of natural origin is quite recent in the industry, but ancient traditions ascribe antioxidant properties to selected plants and herbs in many world areas. The Indian perspective, by the viewpoint of the *Ayurveda*, is extremely interesting. At present, Indian researchers have made several works in the attempt of establishing reliable correlations between antioxidant power and the amount of bio-available polyphenols by herbal extracts. The variety and the large amount of different secondary metabolites may be considered the main reasons for unsatisfactory results when speaking of reliable correlations. On the other side, the use of vegetable polyphenols in the human diet and as a surrogate for synthetic phenolics in food processing should be promoted in the ambit of food safety.

Keywords *Ayurveda* · Food safety · In vitro · In vivo · Oxidation · Phenolics · Secondary metabolite

Abbreviations

AP   Antioxidant Power

© The Author(s), under exclusive license to Springer Nature Switzerland AG 2021   41
R. K. Sharma et al., *Indian Herbal Medicines*, Chemistry of Foods,
https://doi.org/10.1007/978-3-030-80918-8_3

3.1 Introduction to Herbal Extracts as Antioxidant Remedies

The most interesting compounds with claimed and demonstrable antioxidant power concern different molecules in the ambit of natural materials. Generally, an important role is ascribed to phenolics compounds (Haddad et al. 2020a, b; Issaoui et al. 2020; Parisi 2018). Basically, antioxidant power is an interesting argument in the ambit of food and pharma industries, with important implications in the field of cosmetic applications.

In fact, antimicrobial compounds can partially inhibit microbial spreading and related effects, similarly to the inevitability of food perishability, in accordance with the Parisi's First Law of Food Degradation (Anonymous 2021; Baglio 2018; Pellerito et al. 2019; Srivastava 2019). On the other side, antioxidant substances aim at the diminution of damages suffered by organic substrates—foods, beverages, non-food products with a demonstrable perishability or ability to modify their structure and/or superficial appearance during time and living organisms—caused by oxygen attack and consequent oxidative processes. In other terms, each substance able to delay oxidation has to be considered as an antioxidant agent (e.g. polyphenols of vegetable origin). The importance of such a substance is correlated with the decrease of acute and chronic effects of oxidation on organic surfaces, when speaking of organic supports (such as human skin) or food products (Sharma and Rana 2018; Sharma et al. 2019).

The remarkable interest of antioxidant compounds of natural origin is quite recent in the industry, but ancient traditions ascribe antioxidant properties to selected plants and herbs in many world areas. The Indian perspective, by the viewpoint of the *Ayurveda*, is extremely interesting: according to Frawley and Vasant Lad, illnesses are the result of deviations from naturally balanced situations in the human body (Frawley and Lad 1993). As a result, it may be found that a common natural antibiotic as turmeric is cited as a good agent for chronic or only ill patients in the *Ayurveda* ambit and in Louis Pasteur's researches at the same time (Frawley and Lad 1993).

With exclusive reference to natural antioxidants, the Indian researchers have studied the following plants and herbs with the aim of finding some confirmation of claimed antioxidant properties in certain human diseases and in the Ayurvedic ambit (Anusuya et al. 2012; Arunodaya et al. 2016; Ashawat et al. 2007; Ayoub and Mehta 2018; Heer et al. 2016; Gandhiappan and Rengasamy 2012; Kanimozhi and Karthikeyan 2011; Kanimozhi and Prasad, 2009; Ilavarasan et al. 2005; Jagdish et al. 2009; Jamuna et al. 2017; Jayprakash et al. 2002; Lakshmi et al. 2006; Marathe et al. 2011; Mukherjee et al. 2010; Naik et al. 2003; Padma et al. 2001; Ravikumar et al. 2008; Sangeetha et al. 2014; Srinivasan 2014; Suresh and Suriyavathana 2012; Yanpallewar et al. 2004):

(a) *Anisomeles malabarica* (Malabar catmint)
(b) *Annona muricata*

(c)   *Areca catechu* (Betel nut palm)
(d)   *Cassia fistula*
(e)   *Centella asiatica*
(f)   *Cinnamomum tamala* (Indian bay leaf)
(g)   *Curcuma longa* (because of the presence of turmeric oil, obtained as a by-product from the extraction of curcumin)
(h)   *Datura metel*
(i)   *Ganoderma lucidum*
(j)   *Glycirrhiza glabra*
(k)   *Ocimum sanctum*
(l)   *Polyalthia cerasoides*
(m)   *Punica granatum.*

The correlation between the amount of antioxidant molecules—especially phenolics—in herbal extracts and the antioxidant power (AP) is still an issue, at present (Sharma and Rana 2018; Sharma et al. 2019). The main reasons, similarly to studies made on herbal extracts and related antimicrobials, are (Fig. 3.1) as follows:

(1)   The differences between in vivo and in vitro testing methods
(2)   Possible discrepancies related to test protocols.

The main food-related concerns related to natural antioxidants are probably linked to authenticity issues and the possible presence of allergenic substances or reactions (Austin et al. 2018; Casella et al. 2012; Delgado et al. 2016; Dentali et al. 2018;

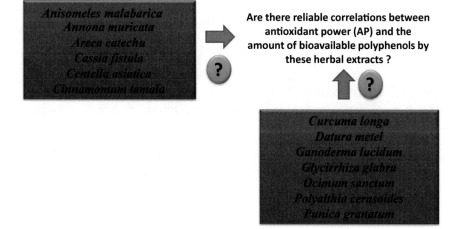

Fig. 3.1 Currently researcher herbal extracts in India with concern to therapeutic uses and Ayurveda

Laganà et al. 2015; Mania et al. 2017; Parisi 2018; Sharma and Parisi 2017; Spanò et al. 2016; Zaccheo et al. 2016).

3.2 Antioxidant Products and Herbal Extracts … in India

Nowadays, polyphenols are considered as a class of specific phytocompounds able to delay oxidative processes. Their presence in vegetable organisms such as *Saraca indica* (*ashoka*), *Lantana camara* (*raimunia*), *G. glabra* (*mulethi*), *Cinnamomum zeylanicum* (*dalchini*), *Acacia arabica* (*babool arabica*) (Katiyar et al. 2018) suggests a good protective effect in humans when patients are able to consume a polyphenols-rich diet. These herbs can be useful against degenerative diseases (Engel et al. 2016; Kalili and de Villiers 2011). Anyway, the subdivision in phenolics classes—phenolic acids; flavonoids (anthocyanidins, chalcones, flavones, flavanols, flavanones, isoflavones, etc.); polyphenolic amides (Tsao 2010)—and the correlated variety of these secondary metabolites can partially explain the natural strategy of defence of plants against ultraviolet radiation and infections (Quideauet al. 2011; Sharma and Rana 2018; Sharma et al. 2019).

With exclusive relation to *Ayurvedic* herbal mixtures and pure herbs, it has been reported that many polyphenols can have interesting but weak therapeutic effects, with the exception of available flavonoids (their prevailing amount in herbs may partially explain their efficacy) (Sharma and Rana 2018; Sharma et al. 2019). The circumstance that potentially healthy foods with antioxidant power can be harmful when ingested in excess and during extended time periods has to be taken into account, according to the Founders of *Ayurveda* (Sharma and Parisi 2017).

3.3 Antioxidant Products and Herbal Extracts in India and Abroad. Conclusions

At present, Indian researchers have made several works in the attempt of establishing reliable correlations between AP and the amount of bio-available polyphenols by herbal extracts. However, the variety and the large amount of different secondary metabolites in the ambit of investigated herbal extracts, including also but not exclusively polyphenols, may be considered the main reasons for unsatisfactory results when speaking of reliable correlations. On the other side, there is little doubt that the use of vegetable polyphenols in the human diet and as a surrogate for synthetic phenolics in food processing should be promoted in the ambit of food safety (Sharma and Rana 2018; Sharma et al. 2019).

References

Anonymous (2021) Parisi's first law of food degradation valuable to establish adequate controls concerning food durability. Inside Lab Manag 21(1):17

Anusuya N, Gomathi R, Manian S, Sivaram V, Menon A (2012 Evaluation of *Basella rubra* L. Rumex nepalensis Spreng. and *Commelina benghalensis* L. for antioxidant activity. Int Res J Pharm Pharm Sci 4(3):714–720

Arunodaya HS, Krishna V, Shashi Kumar R, Kumar KG (2016) Antibacterial and antioxidant activities of stem bark essential oil constituents of *Litsea glutinosa* C.B. Rob. Int J Pharm Pharm Sci 8(12):258–264

Ashawat MS, Shailendra S, Swarnlata S (2007) In vitro antioxidant activity of ethanolic extracts of *Centella asiatica, Punica granatum, Glycyrrhiza glabra* and *Areca catechu*. Res J Med Plant 1(1):13–16. https://doi.org/10.3923/rjmp.2007.13.16

Austin S, van Gool M, Parisi S, Bhandari S, Haselberger P, Jaudzems G, Sullivan D (2018) Standard method performance requirements (SMPRs®) 2014.003: Revised: Determininal of GOS in infant formula and adult/pediatric nutritional formula. J AOAC Int 101(4):1270–1271. https://doi.org/10.5740/jaoacint.SMPR2014.003

Ayoub Z, Mehta A (2018) Medicinal plants as potential source of antioxidant agents: a review. Asian J Pharm Clin Res 11(6):50–56. https://doi.org/10.22159/ajpcr.2018.v11i6.24725

Baglio E (2018) Honey: processing techniques and treatments. In: Chemistry and technology of honey production. Springer International Publishing, Cham

Casella S, Ielati S, Piccione D, Laganà P, Fazio F, Piccione G (2012) Oxidative stress and band 3 protein function in *Liza aurata* and *Salmo irideus* erythrocytes: effect of different aquatic conditions. Cell Biochem Funct 30(5):406–410. https://doi.org/10.1002/cbf.2814

Delgado AM, Vaz Almeida MD, Parisi S (2016) Chemistry of the Mediterranean Diet. Springer International Publishing, Cham, Switzerland. https://doi.org/10.1007/978-3-319-29370-7

Dentali SJ, Amarillas C, Blythe T, Brown PN, Bzhelyansky A, Fields C, Johnson HE, Krepich S, Kuszak A, Metcalfe C, Monagas M, Mudge E, Parisi S, Reif K, Rimmer CA, Sasser M, Solyom AM, Stewart J, Szpylka J, Tims MC, Van Breemen R, You H, Zhao H, Zielinski G, Coates SG (2018) Standard Method Performance Requirements (SMPRs®) 2018.005: Determination of Kavalactones and/or Flavokavains from Kava (*Piper methysticum*). J AOAC Int 101(4):1256–1260. https://doi.org/10.5740/jaoacint.SMPR2018.005

Engel R, Szab K, Abranko LS, Rendes K, Fuzy K, Takacs TN (2016) Effect of *Arbuscular mycorrhizal* fungi on the growth and polyphenol profile of marjoram, lemon balm, and marigold. J Agric Food Chem 64(19):3733–3742. https://doi.org/10.1021/acs.jafc.6b00408

Frawley D, Lad V (eds) (1993) The Yoga of herbs. Lotus Press, Wilmot

Gandhiappan J, Rengasamy R (2012) Comparative study on antioxidant activity of different species of Solanaceae family. Adv Appl Sci Res 3(3):1538–1544

Haddad MA, El-Qudah J, Abu-Romman S, Obeidat M, Iommi C, Jaradat DSM (2020a) Phenolics in Mediterranean and Middle East important fruits. J AOAC Int 103(4):930–934. https://doi.org/10.1093/jaocint/qsz027

Haddad MA, Dmour H, Al-Khazaleh JFM, Obeidat M, Al-Abbadi A, Al-Shadaideh AN, Almazra'awi MS, Shatnawi MA, Iommi C (2020b) Herbs and medicinal plants in Jordan. J AOAC Int 103(4):925–929. https://doi.org/10.1093/jaocint/qsz026

Heer A, Guleria S, Razdan VK (2016) Chemical composition, antioxidant and antimicrobial activities and characterization of bioactive compounds from essential oil of *Cinnamomum tamala* grown in north-western Himalaya. J Plant Biochem Biotechnol 26(2):191–198. https://doi.org/10.1007/s13562-016-0381-7

Ilavarasan R, Mallika M, Venkatraman S (2005) Anti-inflammatory and antioxidant activities of *Cassia fistula* Linn bark extracts. Afr J Trad Compl Altern Med 2(1):70–85. https://doi.org/10.4314/ajtcam.v2i1.31105

Issaoui M, Delgado AM, Caruso G, Micali M, Barbera M, Atrous H, Ouslati A, Chammem N (2020) Phenols, flavors, and the mediterranean diet. J AOAC Int 103(4):915–924. https://doi.org/10.1093/jaocint/qsz018

Jamuna S, Sadullah S, Ashok Kumar R, Shanmuganathan G, Mozhi SS (2017) Potential antioxidant and cytoprotective effects of essential oil extracted from *Cymbopogon citratus* on OxLDL and H_2O_2 LDL induced Human Peripheral Blood Mononuclear Cells (PBMC). Food Sci Hum Wellness 6(2):60–69. https://doi.org/10.1016/j.fshw.2017.02.001

Jayprakasha GK, Jena BS, Negi PS, Sakariah KK (2002) Evaluation of antioxidant activities and antimutagenicity of turmeric oil: a byproduct from curcumin production. Z Naturforsch: C 57(9–10):828–835. https://doi.org/10.1515/znc-2002-9-1013

Kalili KM, de Villiers A (2011) Recent developments in the HPLC separation of phenolic compounds. J Sep Sci 34(8):854–876. https://doi.org/10.1002/jssc.201000811

Kanimozhi P, Karthikeyan J (2011) A study on antioxidant potential of *Glycyrrhiza glabra* linn. In: 1,4-dichlorobenzene induced liver carcinogenesis. J Chem Pharm Res 3(6):288–292

Kanimozhi P, Prasad NR (2009) Antioxidant potential of sesamol and its role on radiation-induced DNA damage in whole-body irradiated Swiss albino mice. Environ Toxicol Pharmacol 28(2):192–197. https://doi.org/10.1016/j.etap.2009.04.003

Katiyar S, Patidar D, Gupta S, Singh RK, Singh P (2018) Some Indian traditional medicinal plants with antioxidant activity: a review. Int J Innov Res Sci Eng Technol 2(12):7421–7432

Laganà P, Melcarne L, Delia S (2015) *Acinetobacter baumannii* and endocarditis, rare complication but important clinical relevance. Int J Cardiol 187:678–679. https://doi.org/10.1016/j.ijcard.2015.04.019

Lakshmi B, Ajith TA, Jose N, Janardhanan KK (2006) Antimutagenic activity of methanolic extract of *Ganoderma lucidum* and its effect on hepatic damage caused by benzo [a] pyrene. J Ethnopharmacol 107(2):297–303. https://doi.org/10.1016/j.jep.2006.03.027

Mania I, Barone C, Pellerito A, Laganà P, Parisi S (2017) Trasparenza e Valorizzazione delle Produzioni Alimentari. L'etichettatura e la Tracciabilità di Filiera come Strumenti di Tutela delle Produzioni Alimentari. Ind Aliment 56(581):18–22

Marathe SA, Rajalakshmi V, Jamdar SN, Sharma A (2011) Comparative study on antioxidant activity of different varieties of commonly consumed legumes in India. Food Chem Toxicol 49(9):2005–2012. https://doi.org/10.1016/j.fct.2011.04.039

Mukherjee M, Bhaskaran N, Srinath R, Shivaprasad HN, Allan JJ, Shekhar D, Agarwal A (2010) Anti-ulcer and antioxidant activity of GutGard. Ind J Exp Biol 48(3):69–274

Naik GH, Priyadarsini KI, Satav JG, Banavalikar MM, Sohoni DP, Biyani MK, Mohan H (2003) Comparative antioxidant activity of individual herbal components used in Ayurvedic medicine. Phytochemistry 63(1):97–104. https://doi.org/10.1016/s0031-9422(02)00754-9

Padma P, Chaurasia JP, Khosa RL, Ray AK (2001) Effect of *Annooa muricata* and *Polyalthia cerasoides* on brain neurotransmitters and enzyme monoamine oxidase following cold immobilization stress. J Nat Remedies 1(2):144–146

Parisi S (2018) Analytical approaches and safety evaluation strategies for antibiotics and antimicrobial agents in food products: chemical and biological solutions. J AOAC Int 101(4):914–915. https://doi.org/10.5740/jaoacint.17-0444

Pellerito A, Dounz-Weigt R, Micali M (2019) Food sharing: chemical evaluation of durable foods. Springer International Publishing, Cham

Quideau S, Deffieux D, Douat-Casassus C, Pouysegu L (2011) Plant polyphenols: chemical properties, biological activities, and synthesis. Angew Chem Int Ed 50(3):586–621. https://doi.org/10.1002/anie.201000044

Ravikumar YS, Mahadevan KM, Kumaraswamy MN, Vaidya VP, Manjunatha H, Kumar V, Satyanarayana ND (2008) Antioxidant, cytotoxic and genotoxic evaluation of alcoholic extract of *Polyalthia cerasoides* (Roxb.) Bedd. Environ Toxicol Pharmacol 26(2):142–146. https://doi.org/10.1016/j.etap.2008.03.001

Sangeetha S, Deepa M, Sugitha N, Mythili S, Sathiavelu A (2014) Antioxidant activity and phytochemical analysis of *Datura metel*. Int J Drug Dev Res 6(4):46–53

Sharma RK, Parisi S (2017) botanical ingredients and herbs in India. Foods or drugs? In: Toxins and contaminants in Indian food products. Springer International Publishing, Cham. https://doi.org/10.1007/978-3-319-48049-7

Sharma RK, Rana BK (2018) Studies on antimicrobial activity and kinetics of inhibition by plant products in India (1990–2016). J AOAC Int 101(4):948–955. https://doi.org/10.5740/jaoacint.17-0449

Sharma RK, Micali M, Pellerito A, Santangelo A, Natalello S, Tulumello R, Singla RK (2019) Studies on the determination of antioxidant activity and phenolic content of plant products in India (2000–2017). J AOAC Int 102(5):1407–1413. https://doi.org/10.1093/jaoac/102.5.1407

Spanò A, Laganà P, Visalli G, Maugeri T, Gugliandolo C (2016) In vitro antibiofilm activity of an exopolysaccharide from the marine thermophilic *Bacillus licheniformis* T14. Curr Microbiol 72(5):518–528. https://doi.org/10.1007/s00284-015-0981-9

Srinivasan K (2014) Antioxidant potential of spices and their active constituents. Crit Rev Food Sci Nutr 54(3):352–372. https://doi.org/10.1080/10408398.2011.585525

Srivastava PK (2019) Status report on bee keeping and honey processing. Development Institute, Ministry of Micro, Small & Medium Enterprises, Government of India 107, Industrial Estate, Kalpi Road, Kanpur-208012, p 64. Available http://msmedikanpur.gov.in/cmdatahien/reports/diffIndustries/Status%20Report%20on%20Bee%20keeping%20&%20Honey%20Processing%202019-2020.pdf. Accessed 10 May 2021

Suresh S, Suriyavathana M (2012) In vitro antioxidant potential of ethanolic extract of *Anisomeles malabarica*. Int Res J Pharm 3(5):394–398

Tsao R (2010) Chemistry and biochemistry of dietary polyphenols. Nutrients 2(12):1231–1246. https://doi.org/10.3390/nu2121231

Yanpallewar SU, Rai S, Kumar M, Acharya SB (2004) Evaluation of antioxidant and neuroprotective effect of *Ocimum sanctum* on transient cerebral ischemia and long-term cerebral hypoperfusion. Pharmacol Biochem Behav 79(1):155–164. https://doi.org/10.1016/j.pbb.2004.07.008

Zaccheo A, Palmaccio E, Venable M, Locarnini-Sciaroni I, Parisi S (2016) Food hygiene and applied food microbiology in an anthropological cross cultural perspective. Springer International Publishing, Cham, Switzerland. https://doi.org/10.1007/978-3-319-44975-3

Chapter 4
Natural Antioxidant Agents for Treatment of Metabolic Diseases and Disorders

Abstract This chapter concerns the possible use of natural molecules from vegetable sources in food and pharma preparations, including also the ambit of cosmetic industries. This possibility has been often considered with success in the recent years because of 'natural' claimed properties. In general, polyphenols and other substances named as 'natural' additives or nutraceuticals are considered and advertised as good surrogates for synthetic additives because of the following claimed feature at least: antioxidant power. In this ambit, the main part of natural antioxidant agents of vegetable origin is secondary metabolites responsible for different plant features and—consequently—food features such as colours and aroma: Phenolics; terpenoids; some proteins; chain-breakers or scavengers and, last but not least, vegetable enzymes and several mineral elements. Because of the relationship between oxidative stress (cause) and pathophysiological effects (illness) on the human health, the use of new and non-synthetic compounds with antioxidant power should be preferred and recommended in the ambit of well-balanced dietary lifestyles. Consequently, therapeutic or economic accessibility to antioxidants may be difficult, unless vegetable plants and herbal extracts are considered as strong antioxidant foods. In conclusion, the consumption of antioxidant compounds from readily and easily available foods of vegetable origin should be recommended when speaking of public health issues, and the promotion of traditional medicine may explain partially interesting medicinal effects in this ambit.

Keywords Antioxidant power · Illness · Natural · Oxidation · Phenolics · Public health · Traditional medicine

Abbreviations

AP   Antioxidant Power
MDD   Metabolic Diseases and Disorder
ROS   Reactive Oxygen Species
RNS   Reactive Nitrogen Species

© The Author(s), under exclusive license to Springer Nature Switzerland AG 2021   49
R. K. Sharma et al., *Indian Herbal Medicines*, Chemistry of Foods,
https://doi.org/10.1007/978-3-030-80918-8_4

4.1 Natural Antioxidants Today. Why?

The possible use of natural molecules from vegetable sources in food and pharma preparations, including also the ambit of cosmetic industries, has been often considered with success in the recent years because of 'natural' claimed properties. In general, polyphenols and other substances named as 'natural' additives or nutraceuticals are considered and advertised as good surrogates for synthetic additives because of the following claimed features:

(a) Antioxidant power (Issaoui et al. 2020; Laganà et al. 2019; Parisi 2020)
(b) Other properties able to contrast and prevent metabolic diseases and disorders (MDD), including cardiovascular diseases, prevention of chronic degenerative diseases (De et al. 2020; Scotti et al. 2016; Shen and Singla 2020; Singla 2020; Singla et al. 2018, 2020).

In this ambit, the main part of natural antioxidant agents of vegetable origin is secondary metabolites responsible for different plant features and—consequently—food features such as colours and aroma (Fig. 4.1):

(1) Phenolics compounds. For example: Eugenol, ferulic acid, rosmarinic acid, naringenin, caffeic acid, punicalagin, aloesin, etc. Phenolics are considered the main part of naturally occurring antioxidants from vegetable sources (Arulselvan et al. 2016; Batista 2014; Haddad et al. 2020a, b; Imran et al. 2020; Martins Gregório et al. 2016)

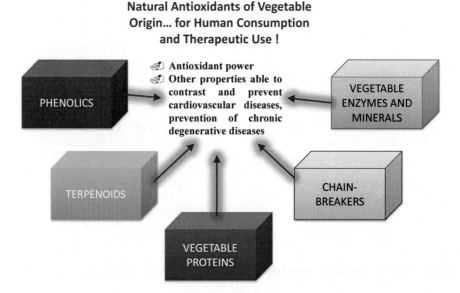

Fig. 4.1 Natural antioxidants of vegetable origins. A simple classification

(2) Terpenoids. For example: Lycopene; several monoterpenes (α-terpineol α-thujone, R-(+)- α -pinene, α-pinene (racemic mixture), α-terpinene, limonene, p-cymene, linalool, etc.) (Bicas et al. 2011; de Oliveira et al. 2015; Mothana et al. 2001; Ruberto and Baratta 2000; Wannes et al. 2010)

(3) Some proteins. For example: Albumin, caeruloplasmin, transferring, etc. (Maxwell 1995)

(4) 'Chain-breakers' or scavengers: Uric acid, bilirubin, ascorbic acid, tocopherol (vitamin E) and β-carotene (a terpenoid) (Maxwell 1995)

(5) Vegetable enzymes and several mineral elements (copper, magnesium, manganese, selenium, etc.) (Apel et al. 2001; Arulselvan et al. 2016).

It has to be explained that antioxidant compounds and/or their mixtures are recognised to delay and prevent oxidative reactions (also named 'oxidative stress') in living life forms, such as the human being. In particular, and with relation to humans, oxidative stress is caused in tissues by free radicals (lipid peroxyl, alkoxyl and peroxide; hydroxyl free radical), reactive oxygen species (ROS) such as hydrogen peroxide and reactive nitrogen species (RNS) (Aruselvan et al. 2016). The result of abnormal amounts of these species, generally enzymes (e.g. catalase, glutathione peroxidase and superoxide dismutase), is the prevailing oxidising action with demolition of tissues and cellular damages (Costa Junior et al. 2011; Wickens 2001). Because of the relationship between oxidative stress (cause) and pathophysiological effects (illness) on the human health (Santos et al. 2011), the use of new and non-synthetic compounds with antioxidant power (AP) should be preferred and recommended in the ambit of well-balanced dietary lifestyles. The common point for all of these substances is the antioxidant action, while several of these compounds belong to the Animal Kingdom, and other compounds may be of synthetic origin. Consequently, therapeutic or economic accessibility to antioxidants may be difficult, unless vegetable plants and herbal extracts are considered as high-AP foods.

As a result,

(1) The presence of high-AP compounds in vegetable foods (tomatoes, pomegranate, olives, olive oil, onions and leaves/roots of other plants) has to be taken into account when speaking of important pillars in the definition of selected lifestyles such as Mediterranean diet (Delgado et al. 2016; Issaoui et al. 2020).

(2) In addition, the following diseases of the human being—atherosclerosis; ischemia; diabetes mellitus; Alzheimer's disease; acute and chronic inflammations; hypertension; cancers; rheumatoid arthritis; acute cerebral stroke—can be contrasted with interesting results when high-AP foods of vegetable origin are available. The World Health Organization has estimated that 4/5 of the world population were accustomed to use traditional medicine and associated remedies for primary health issues (Aruselvan et al. 2016; Maxwell 1995).

4.2   Natural Antioxidants Today. Conclusions and Perspectives

In conclusion, the consumption of antioxidant compounds from readily and easily available foods of vegetable origin should be recommended when speaking of public health issues, and the promotion of traditional medicine may explain partially interesting medicinal effects in this ambit (Campêlo 2011; Costa Junior et al. 2011; Freitas 2001, 2009; Hertog et al. 1993; Maciel et al. 2002; Mengues et al. 2011; Nogueira Neto et al. 2013; Veiga Junior et al. 2005). Antioxidant molecules of natural origin have to be consumed in the right amount, without excess. Because of the possibility of adverse effects on humans caused by gastrointestinal side effects and other collateral (undesired) problems when speaking of consumption of synthetic drugs or analgesics, the use of natural remedies may be useful (Albarracin et al. 2012; Balsano and Alisi 2009; Choi et al. 2012; Frei 2012; Kaur and Kapoor 2002; Podsędek 2007; Rafieian-Kopaei et al. 2013; Wang et al. 2011). Moreover, it is predictable that the availability of plants and herbal extracts and the cheapness of these foods can be good factors in the treatment of metabolic diseases and disorders.

References

Albarracin SL, Stab B, Casas Z, Sutachan JJ, Samudio I, Gonzalez J, Capani F, Morales L, Barreto GE (2012) Effects of natural antioxidants in neurodegenerative disease. Nutr Neurosci 15(1):1–9. https://doi.org/10.1179/1476830511Y.0000000028

Apel M, Limberger RP, Sobral M, Menut C, Henriques A (2001) Chemical composition of the essential oil of *Siphoneugena reitzii* D. Legr. J Essent Oil Res 13(6):429–430. https://doi.org/10.1080/10412905.2001.9699716

Arulselvan P, Fard MT, Tan WS, Gothai S, Fakurazi S, Norhaizan ME, Kumar SS (2016) Role of antioxidants and natural products in inflammation. Oxid Med Cell Longev 2016:Article ID 5276130. https://doi.org/10.1155/2016/5276130

Balsano C, Alisi A (2009) Antioxidant effects of natural bioactive compounds. Curr Pharm Des 15(26):3063–3073. https://doi.org/10.2174/138161209789058084

Batista R (2014) Uses and potential applications of ferulic acid. Warren B (ed) Ferulic acid: antioxidant properties, uses and potential health benefits, 1st edn. Nova Science Publishers Inc., New York, pp 39–70

Bicas JL, Neri-Numa IA, Ruiz ALTG, De Carvalho JE, Pastore GM (2011) Evaluation of the antioxidant and antiproliferative potential of bioflavors. Food Chem Toxicol 49(7):1610–1615. https://doi.org/10.1016/j.fct.2011.04.012

Campêlo LML (2011) Pharmacological evaluation of the essential oil of *Citrus limon* (Burm) central nervous system: a study behavioral, neurochemical and histological. Dissertation, Federal University of Piauí, Teresina, Brazil

Choi DY, Lee YJ, Hong JT, Lee HJ (2012) Antioxidant properties of natural polyphenols and their therapeutic potentials for Alzheimer's disease. Brain Res Bull 87(2–3):144–153. https://doi.org/10.1016/j.brainresbull.2011.11.014

Costa Junior JS, Ferraz ABF, Filho BAB, Feitosa CM, Citó AMGL, Freitas RM, Saffi J (2011) Evaluation of antioxidant effects in vitro of garcinielliptone FC (GFC) isolated from *Platonia insignis* Mart. J Med Plants Res 5(2):293–299

De B, Bhandari K, Singla RK, Saha G, Goswami TK (2020) In silico molecular GRIP docking of some secondary metabolites combating diabesity. Bull Natl Res Centre 44(1):1–13. https://doi.org/10.1186/s42269-020-00327-7

de Oliveira TM, de Carvalho RBF, da Costa IHF, de Oliveira GAL, de Souza AA, de Lima SG, de Freitas RM (2015) Evaluation of p-cymene, a natural antioxidant. Pharm Biol 53(3):423–428. https://doi.org/10.3109/13880209.2014.923003

Delgado AM, Vaz Almeida MD, Parisi S (2016) Chemistry of the Mediterranean Diet. Springer International Publishing, Cham, Switzerland. https://doi.org/10.1007/978-3-319-29370-7

Frei B (ed) (2012) Natural antioxidants in human health and disease. Academic Press, San Diego

Freitas RM (2001) Antioxidant properties and therapeutic potential of essential oils. In: Sousa DP (ed) Medicinal essential oils: chemical, pharmacological and therapeutic aspects. New Science Publishers, New York, pp 1–9

Freitas RM (2009) The evaluation of effects of lipoic acid on the lipid peroxidation, nitrite formation and antioxidant enzymes in the hippocampus of rats after pilocarpine-induced seizures. Neurosci Lett 455(2):140–144. https://doi.org/10.1016/j.neulet.2009.03.065

Haddad MA, Dmour H, Al-Khazaleh JFM, Obeidat M, Al-Abbadi A, Al-Shadaideh AN, Al-mazra'awi MS, Shatnawi MA, Iommi C (2020) Herbs and medicinal plants in Jordan. J AOAC Int 103(4):925–929. https://doi.org/10.1093/jaocint/qsz026

Haddad MA, El-Qudah J, Abu-Romman S, Obeidat M, Iommi C, Jaradat DSM (2020) Phenolics in Mediterranean and Middle East important fruits. J AOAC Int 103(4):930–934. https://doi.org/10.1093/jaocint/qsz027

Hertog MG, Feskens EJ, Kromhout D, Hollman PCH, Katan MB (1993) Dietary antioxidant flavonoids and risk of coronary heart disease: the Zutphen Elderly Study. Lancet 342(8878):1007–1011. https://doi.org/10.1016/0140-6736(93)92876-U

Imran M, Ghorat F, Ul-Haq I, Ur-Rehman H, Aslam F, Heydari M, Shariati MA, Okuskhanova E, Yessimbekov Z, Thiruvengadam M, Hashempur MA, Rebezov M (2020) Lycopene as a natural antioxidant used to prevent human health disorders. Antioxid 9(8):706. https://doi.org/10.3390/antiox9080706

Issaoui M, Delgado AM, Caruso G, Micali M, Barbera M, Atrous H, Ouslati A, Chammem N (2020) Phenols, flavors, and the Mediterranean diet. J AOAC Int 103(4):915–924. https://doi.org/10.1093/jaocint/qsz018

Kaur C, Kapoor HC (2002) Anti-oxidant activity and total phenolic content of some Asian vegetables. Int J Food Sci Technol 37(2):153–161. https://doi.org/10.1046/j.1365-2621.2002.00552.x

Laganà P, Anastasi G, Marano F, Piccione S, Singla RK, Dubey AK, Coniglio MA, Facciolà A, Di Pietro A, Haddad MA, Al-Hiary M, Caruso G (2019) Phenolic substances in foods: health effects as anti-inflammatory and antimicrobial agents. J AOAC Int 102(5):1378–1387. https://doi.org/10.1093/jaoac/102.5.1378

Maciel MAM, Pinto AC, Veiga Junior VF, Grynberg NF, Echevarria A (2002) Medicinal plants: the need for multidisciplinary studies. New Chem 25(3):429–438. https://doi.org/10.1590/S0100-40422002000300016

Martins Gregório B, Benchimol De Souza D, de Morais A, Nascimento F, Matta L, Fernandes-Santos C (2016) The potential role of antioxidants in metabolic syndrome. Curr Pharm Des 22(7):859–869. https://doi.org/10.2174/1381612822666151209152352

Maxwell SR (1995) Prospects for the use of antioxidant therapies. Drugs 49(3):345–361. https://doi.org/10.2165/00003495-199549030-00003

Mengues SS, Mentz LA, Schenkel EP (2011) Use of medicinal plants in pregnancy. Rev Bras Farmacogn 11:21–35

Mothana RAA, Hasson SS, Schultze W, Mowitz A, Lindequist U (2001) Phytochemical composition and in vitro antimicrobial and antioxidant activities of essential oils of three endemic Soqotraen boswellia species. Food Chem 126:1149–1154

Nogueira Neto JD, Sousa DP, Freitas RM (2013) Evaluation of antioxidant potential in vitro of nerolidol. J Pharmaceut Sci Basic Appl 34:125–130

Parisi S (2020) Characterization of major phenolic compounds in selected foods by the technological and health promotion viewpoints. J AOAC Int 103(4):904–905. https://doi.org/10.1093/jaoacint/qsaa011

Podsędek A (2007) Natural antioxidants and antioxidant capacity of Brassica vegetables: a review. LWT-Food Sci Technol 40(1):1–11. https://doi.org/10.1016/j.lwt.2005.07.023

Rafieian-Kopaei M, Baradaran A, Rafieian M (2013) Plants anti-oxidants: from laboratory to clinic. J Nephropathol 2(2):152–153. https://doi.org/10.12860/JNP.2013.26

Ruberto G, Baratta MT (2000) Antioxidant activity of selected essential oil components in two lipid model systems. Food Chem 69:167–174. https://doi.org/10.1016/S0308-8146(99)00247-2

Santos PS, Costa JP, Tome´ AR, Saldanha GB, de Souza GF, Feng D, de Freitas RM (2011) Oxidative stress in rat striatum af-ter pilocarpine-induced seizures is diminished by alphatocopherol. Eur J Pharmacol 668(1–2):65–71.https://doi.org/10.1016/j.ejphar.2011.06.035

Scotti L, Singla RK, Scotti MT (2016) Editorial (thematic issue: natural leads in drug discovery against metabolic disorders and their related infectious diseases). Curr Top Med Chem 16(23):2523–2524. https://doi.org/10.2174/1568026616999160510121418

Shen B, Singla RK (2020) Secondary metabolites as treatment of choice for metabolic disorders and infectious diseases & their metabolic profiling-Part 2. Curr Drug Metab 21(14):1070–1071. https://doi.org/10.2174/138920022114201230142204

Shen L, Shen K, Bai J, Wang J, Singla RK, Shen B (2020) Data-driven microbiota biomarker discovery for personalized drug therapy of cardiovascular disease. Pharmacol Res 161:105225. https://doi.org/10.1016/j.phrs.2020.105225

Singla RK (2020) Secondary metabolites as treatment of choice for metabolic disorders and infectious diseases and their metabolic profiling: Part 1. Curr Drug Metab 21(7):480–481. https://doi.org/10.2174/1389200221072009251016311

Singla RK, Ali M, Kamal MA, Dubey AK (2018) Isolation and characterization of nuciferoic acid, a novel keto fatty acid with hyaluronidase inhibitory activity from *Cocos nucifera* Linn. Endocarp. Curr Top Med Chem 18(27):2367–2378. https://doi.org/10.2174/1568026619666181224111319

Veiga Junior VF, Pinto AC, Maciel MAM (2005) Medicinal plants: safe cure? New Chem 28(3):519–528. https://doi.org/10.1590/S0100-40422005000300026

Wang S, Melnyk JP, Tsao R, Marcone MF (2011) How natural dietary antioxidants in fruits, vegetables and legumes promote vascular health. Food Res Int 44(1):14–22. https://doi.org/10.1016/j.foodres.2010.09.028

Wannes WA, Mhamdi B, Sriti J, Jemia MB, Ouchikh O, Hamdaoui G, Kchouk ME, Marzouk B (2010) Antioxidant activities of the essential oils and methanol extracts from myrtle (*Myrtus communis* var. italica L.) leaf, stem and flower. Food Chem Toxicol 48(5):1362–1370. https://doi.org/10.1016/j.fct.2010.03.002

Wickens AP (2001) Ageing and the free radical theory. Respir Physiol 128(3):379–391. https://doi.org/10.1016/S0034-5687(01)00313-9

Printed in the United States
by Baker & Taylor Publisher Services